PSYCHIATRIC REHABILITATION

William Anthony
Mikal Cohen
Marianne Farkas

Psychiatric Rehabilitation

Center for
Psychiatric
Rehabilitation

Boston University
Sargent College of Allied Health Professions

Boston

Center for Psychiatric Rehabilitation
730 Commonwealth Avenue
Boston, Massachusetts 02215

Printed in the United States of America

Book design by Linda Getgen
Copyediting and indexing by Wordsworth Associates
Typesetting, printing, and binding by Arcata Graphics

Library of Congress Catalog Card Number: 89-062415

ISBN 1-878512-00-5

To Robert Carkhuff—our mentor

No royalties are paid to the authors of this text; all proceeds from the sale of this text will go to further the work of the Center for Psychiatric Rehabilitation.

Contents

Foreword

In its groundbreaking report to the nation, *Action for Mental Health*, the Joint Commission on Mental Illness and Health (1961) wrote:

> Events of recent years have demonstrated unequivocally the value of positive efforts toward rehabilitation of patients who have been mentally ill, or have had chronic neurological diseases. Examination of our methods in rehabilitation, however, would indicate that they are crude, imprecise, and not highly specific for persons with different kinds of mental and character disorders. Much research is needed in this area. Further inquiry is necessary into the kinds of persons who make the best rehabilitation workers and the kinds of professional skills that should be incorporated into this new professional group. While there is a heady interest in rehabilitation and such services have widespread popular acceptance, sufficient support for research in the area is an urgent requirement.

Psychiatric Rehabilitation accepts the challenge that was laid down three decades ago in this report. It examines in detail the state of the art of psychiatric rehabilitation and assesses its technological base, its research foundation, and its personnel requirements, as called for by the Joint Commission.

Some might think this an unsatisfactorily long time from charge to fulfillment, but a careful reading of *Psychiatric Rehabilitation* suggests that evolution in this field has necessarily been slow. Complex problems, after all, require complex and carefully crafted solutions. What could be more complicated, and more needful of cautious advance, than the tasks of conceptualizing services for a diverse population whose elements have for too long been lumped together under the heading of "the mentally ill"?

Indeed, *Psychiatric Rehabilitation* suggests that the field is still evolving. This book takes a cross-sectional view of a dynamic and changing discipline and provides us with a long-overdue progress report. As noted here, the approach developed by Boston University, where the authors are affiliated, is one of several current approaches to psychiatric rehabilitation, and its founders are widely lauded as pioneers in this area.

This volume appears at a critical time in the history of mental health service planning and service delivery. The demography, service histories, and program needs of individuals with severe and disabling mental illness have changed dramatically over the past three decades, as fewer patients spend long periods of time in institutional settings. Increasing numbers of mentally ill persons never enter institutions at all and receive all of their care, whatever its content, in the community. New service structures and innovative approaches to care are continually being tested. Some have been discarded; others have established themselves as part of the growing armamentarium of service interventions. Throughout all, the objectives of improving the quality of life for mentally ill individuals and of helping them realize their full potential as human beings have persisted, but the changes have often raised more questions than they have answered. They have certainly generated a series of paradoxes.

For example, on the positive side, we have learned to think about the needs of mentally ill persons in new ways, a major and, we trust, lasting advance. There is increasing recognition of the importance of social support systems, of involving families in the care of their disabled relatives, of planning creative residential alternatives, and of refining rehabilitation techniques, as more providers acknowledge the need to treat both illness and disability.

Counterbalancing these positive developments, however, is the reality that existing service structures and service systems do not lend themselves readily to providing continuity in the care of mentally ill individuals. Similarly, the fragmentation of services and authority has limited the ability of many programs to be comprehensive in their scope. And a serious—and in many areas growing—scarcity of resources has often placed mentally ill individuals in the position of having to compete for services and prove their eligibility. The notion of service entitlement for disabled individuals is a very precarious one in American culture.

A realistic appraisal of the politics of mental health service delivery suggests that these problems may intensify in years to come. As programs become more comprehensive in their scope, and as the individuals served in them become more diverse in their treatment histories, we shall require more complex networks of services and more sophisticated administrative techniques. We shall need more and better-distributed service structures to respond to the multiplicity of service needs experienced by mentally ill individuals. Above all, we shall need professionals who are committed to the goals of humane and sensitive care for the members of this population.

These are not easy goals to pursue, but *Psychiatric Rehabilitation*, with its hopeful and positive message, encourages optimism. It is a book that calls to mind the words of Hubert Humphrey (1977) on the occasion of the dedication of the federal Department of Health and Human Service building that bears his name. Seriously ill with cancer at that time, Humphrey noted, "There's a great deal of difference between failure and defeat. Failure means giving up, you're through. Defeat means you wait for another chance to come back, to fight the good battle."

In *Psychiatric Rehabilitation* the approach to the care of mentally ill individuals encourages them, as well as the professionals who work with them, to "fight the good battle." It is one of a complex of approaches needed to serve the diverse and heterogeneous population of mentally ill persons, and, as the

xiii

authors ably demonstrate, it intersects with more traditional medical and clinical interventions at many junctures. It is an approach that has been painstakingly developed and refined and that is presented to us here in detail.

Leona L. Bachrach, Ph.D.

References

Humphrey, H. H. (1977, November 4). Untitled remarks. Congressional Record, Vol. 123, p. 37287.

Joint Commission on Mental Illness and Health. (1961). Action for Mental Health. New York: Basic Books, pp. 274–275.

Preface

Since the publication in 1979 of *The Principles of Psychiatric Rehabilitation* and the six-volume *Psychiatric Rehabilitation Practice Series*, the Center for Psychiatric Rehabilitation at Boston University has been deeply involved in demonstrating that the field of psychiatric rehabilitation is a necessary, rewarding, and credible field of study.

At the beginning of the 1980s the Center was part of a psychiatric rehabilitation field that had not achieved consensus on its underlying philosophy, had not integrated its research studies into a substantial knowledge base, had not widely publicized its model service programs, and had not articulated a rehabilitation technology. Throughout the Center's first decade, the Center's research, training, and direct service were aimed at working with other persons in the field to overcome these deficiencies.

At the beginning of the 1990s considerable agreement exists on the fundamental philosophy of psychiatric rehabilitation, a significant body of research forms the knowledge base of a credible psychiatric rehabilitation field, a variety of model service programs has been demonstrated and disseminated, preservice and inservice professional training programs are now in place, and a technology exists.

The Center has specifically contributed to these advances by:

- Writing articles, book chapters, and books that organize the concepts and empirical data relevant to psychiatric rehabilitation.
- Conducting studies researching the impact of various psychiatric rehabilitation interventions on persons with psychiatric disabilities.
- Presenting at workshops and conferences around the world.
- Disseminating information through a computerized data base service, an electronic bulletin board service, and jointly editing and publishing an international journal (*Psychosocial Rehabilitation Journal*) and newsletter (*Community Support Network News*).
- Developing, implementing, and evaluating model programs of supported living, supported learning, and supported employment.
- Conducting inservice training and program consultation throughout North America and Europe.
- Educating bachelor's, master's, and doctoral students in a state-of-the-art psychiatric rehabilitation curriculum at Boston University.
- Developing a training technology that has been used to train large numbers of psychiatric rehabilitation practitioners.

This book pulls together what staff at the Center for Psychiatric Rehabilitation have learned during the last ten years. It draws from published and unpublished material and represents our latest thinking on the history, current status, and future direction of the psychiatric rehabilitation field.

The book is written as an introduction to the field for those who wish to study and specialize in psychiatric rehabilitation, and as an overview for those who are involved with the field of psychiatric rehabilitation because of their role as administrator, practitioner, researcher, family member, or consumer advocate. The text equips the reader with information about the field's conceptual, research, and technological base. The extensive and current bibliography of the major articles, book chapters, and books can provide the reader with sources of additional information about each topic area.

Staff from the Center for Psychiatric Rehabilitation worked in the decade of the 1970s in a field still reeling from the aftereffects of deinstitutionalization. The 1980s can be seen as a decade of transitions—a transition between the former era of deinstitutionalization and the future era of rehabilitation. Staff at the Center for Psychiatric Rehabilitation believe that the decade of the 1990s is the beginning of the rehabilitation era. It is toward that end that this book is written.

Acknowledgments

First and foremost, we would like to acknowledge the significant contributions of those persons with psychiatric disabilities and their family members who work with the Center for Psychiatric Rehabilitation. They have helped us to grow personally and professionally.

Additionally, we would like to acknowledge the enormous contributions of our colleagues at the Center for Psychiatric Rehabilitation. Every Center staff member has contributed in some way to the writing of this book. Support staff guided by Aurora Wilber-Leach and Jennifer Tratnyek patiently and competently transformed our fevered scribbles to intelligible narrative. Direct service staff—Larry Kohn, Dori Hutchinson, Larry Ward, Dale Walsh, Laura Mancuso, Ann Sullivan, and Kim McDonald-Wilson—led by LeRoy Spaniol have shared their understanding of the needs and wants of their clients and insights about the rehabilitation services of the future. Karen Unger, Karen Danley, and Priscilla Ridgway have helped to conceptualize psychiatric rehabilitation applications in innovative settings. The research staff—Sally Rogers, Ken Sciarappa, and Yumi Williston—have shared their knowledge of research studies and reinforced our appreciation of the importance of credible research. The technology dissemination staff—Bill Kennard, Rick Forbess, Bill O'Brien, Dean Mynks, Pat Nemec, Judith Taylor, and Sue McNamara—have shared their contagious commitment to the use of psychiatric rehabilitation technology based on experiences in hundreds of service settings and have helped develop our vision of a client-driven service system. Training staff—Kathy Furlong-Norman and Susan Hecht—have helped us stay connected to the concerns of the field through their patient compilation and dissemination of community support and rehabilitation information. Judi Chamberlin and other ex-patient staff have shared with us their unique perspectives on psychiatric rehabilitation and have helped us to reflect on the real implications of our work. Barry Cohen has contributed to the articulation of the philosophy, process, and technology of psychiatric rehabilitation. His insights and conceptual organization are strongly reflected in the approach presented in this book.

We owe a special thank you to Linda Getgen. She used her artistic talent, production expertise, and negotiating skills to produce ''the book'' smoothly.

William Anthony
Mikal Cohen
Marianne Farkas

1

Introduction

How many times it thundered before Franklin took the hint! How many apples fell on Newton's head before he took the hint! Nature is always hinting at us. It hints over and over again. And suddenly we take the hint.

Robert Frost

The essential elements of a psychiatric rehabilitation approach have been hinted at for well over a century. Different elements of a psychiatric rehabilitation approach have periodically moved in and out of favor, highlighted almost serendipitously as the mental health field progressed through various developmental phases.

The 1980s has been a decade of transitions—a transition between the former era of deinstitutionalization and the future era of rehabilitation. The 1980s sounded the death knell for whatever was left of the deinstitutionalization era while at the same time ushering in the era of rehabilitation. The decade of the 1990s may be seen as the decade when psychiatric rehabilitation will assume its rightful place as one of the triumvirate of mental illness initiatives, that is, prevention, treatment, and rehabilitation.

For more than three decades deinstitutionalization issues have predominated in the professional journals and the popular press, in contrast to the limited space devoted to rehabilitation issues. Now, however, that preoccupation seems to be shifting to a new interest in rehabilitation.

Deinstitutionalization and rehabilitation are significantly

1

different. Deinstitutionalization focused on how buildings function; rehabilitation focuses on how people function. Deinstitutionalization focused on closing buildings; rehabilitation focuses on opening lives. Deinstitutionalization focused on getting rid of patient restraints; rehabilitation focuses on getting personal supports.

In contrast to the deinstitutionalization initiative, with its focus on emptying buildings, it is easy to be excited and enthusiastic about rehabilitation, with its focus on improved quality of life. Interestingly, a deinstitutionalization goal of fewer people living in institutions for fewer days can also be realized by rehabilitation. However, rehabilitation values and principles guide practitioners in achieving this same outcome.

In the final analysis, deinstitutionalization, because it was done so poorly, was a relatively easy task in comparison to rehabilitation. Deinstitutionalization opened the doors of the hospital and gave people a prescription for their medicine when they left. Rehabilitation tries to open the doors of the community and help people develop a prescription for their lives. The deinstitutionalization era is yesterday's focus. The rehabilitation era is upon us. It guides our current activities and gives us a vision of the future. Deinstitutionalization is now a historical fact. A return to the era that preceded it, that is, to institutionalization of large numbers of persons with psychiatric disabilities, is an economic impossibility. Society will simply not pay for it. The question is, Where do we go from here? The answer to that question is provided by psychiatric rehabilitation.

The Field of Psychiatric Rehabilitation

The developing consensus about the overall mission of psychiatric rehabilitation is: Psychiatric rehabilitation assists persons with long-term psychiatric disabilities increase their functioning so that they are successful and satisfied in the environments of their choice with the least amount of ongoing professional intervention (Anthony, Cohen, & Cohen, 1983). The major methods by which this mission is accomplished involve either developing the specific skills the client needs to function effectively and/ or developing the resources needed to support or strengthen the

2

client's present levels of functioning (Anthony, 1979; Anthony et al., 1983; Liberman & Fox, 1983).

The term *psychiatric rehabilitation* is becoming routinely used in the mental health field, increasingly in both treatment professionals' jargon and administrators' program descriptions. Within the past decade, psychiatric rehabilitation has begun to take its place as a viable, credible service.

The term *psychiatric rehabilitation* has become so overused that it has become necessary to say both what it is and what it is not. The word *psychiatric* was selected to describe the disability that is the focus of the rehabilitation. This does not mean that treatment must be done by psychiatrists or using psychiatric treatment methods. The term *rehabilitation* reflects the focus of the approach on improved functioning in a specific environment. Although many different techniques and settings are used in the rehabilitation of persons with psychiatric disabilities (e.g., social skills training, clubhouses), the field of rehabilitation shares a common philosophy.

The work of current researchers and practitioners will determine whether psychiatric rehabilitation remains a viable, credible field of study and practice, or merely a historical footnote. At present, many mental health professionals now recognize the need for a rehabilitation approach to complement existing treatment approaches (Anthony, 1977; Liberman & Foy, 1983). However, this recognition of need does not mean that psychiatric rehabilitation is well understood. Because all types of mental health disciplines practice psychiatric rehabilitation, and because relevant research and conceptual articles appear in a wide range of professional journals, psychiatric rehabilitation has been, until recently, difficult to define and understand.

Psychiatric Rehabilitation is an attempt to meet this need. In essence, this book examines the current state of the field and suggests future directions. It traces the history of psychiatric rehabilitation in terms of relevant historical developments and discarded myths. The current status of psychiatric rehabilitation is overviewed in terms of its research base, its conceptual foundation, underlying philosophy, technology, and existing practice.

A highlight is the emerging technology of psychiatric rehabilitation, which permits the comprehensive training of practitioners, the monitoring and evaluation of practice, the development and replication of programs, the empirical investigation of the essential ingredients of psychiatric rehabilitation, and the integration of a comprehensive psychiatric rehabilitation approach into mental health service systems.

The Persons in Need of Psychiatric Rehabilitation

Psychiatric rehabilitation focuses on persons who have experienced severe psychiatric disabilities rather than on individuals who are simply dissatisfied, unhappy, or "socially disadvantaged." Persons with severe psychiatric disability have diagnosed mental illnesses that limit their capacity to perform certain functions (e.g., conversing with family and friends, interviewing for a job) and their ability to perform in certain roles (e.g., worker, student).

Several attempts at defining severe psychiatric disability present a convergent description of this disability. Three of these definitions are the definitions currently used by the Community Support Program of the National Institute of Mental Health (NIMH), the Rehabilitation Service Administration's (RSA) definition of severe disability, and Goldman's (Goldman, Gattozzi, & Taube, 1981) definition of the "chronically mentally ill."

The Community Support Program (CSP) has identified the "chronically mentally ill" as its target population. In this book the term *chronically mentally ill* or its abbreviation—CMI—is not used because of the stigmatizing label and pessimistic expectations connoted by this term. Based on several years of CSP projects, NIMH developed an operational definition of adult clients served by the CSP initiative (National Institute of Mental Health, 1980). This definition includes important characteristics of the population that is the focus of this book (see Table 1–1).

Another working definition of this population has been provided by Goldman et al. (1981). They describe the severely psychiatrically disabled in terms of diagnosis, disability, and duration. The estimation of from 1.7 to 2.4 million persons with severe psychiatric disabilities is often based on their analysis. Goldman et al. (1981) have identified the "chronically mentally ill" as having a severe mental disorder (e.g., typically psychosis) with moderate

4

TABLE 1–1 *The Community Support Program (CSP): Definition of Target Population*

1. Severe Disability Resulting from Mental Illness

CSP clients typically meet at least *one* of the following criteria:

- Have undergone psychiatric treatment more intensive than outpatient care more than once in a lifetime (e.g., emergency services, alternative home care, partial hospitalization, or inpatient hospitalization).

- Have experienced a single episode of continuous, structured, supportive residential care other than hospitalization for a duration of at least two years.

2. Impaired Role Functioning

CSP clients typically meet at least *two* of the following criteria on a continuing or intermittent basis for at least two years.

- Are unemployed, are employed in a sheltered setting, or have markedly limited skills and a poor work history.

- Require public financial assistance for out-of-hospital maintenance and may be unable to procure such assistance without help.

- Show severe inability to establish or maintain a personal social support system.

- Require help in basic living skills.

- Exhibit inappropriate social behavior, which results in demand for intervention by the mental health and/or judicial system.

Adapted from: National Institute of Mental Health. (1980). *Announcement of community support system strategy development and implementation grants* (pp. iii, iv). Rockville, MD: Author.

to severe disability (e.g., functional incapacity) of prolonged duration (e.g., a period of supervised residential care).

Another definition of severity is used by RSA as described by the 1973 Rehabilitation Act. The act defines a severe handicap as a "disability which requires multiple services over an extended period of time." One specific impairment cited as causing a severe disability is mental illness. A disability is defined as either a physical or mental condition that limits a person's activities or functioning. The severely disabled are the first priority of service as mandated by the Rehabilitation Act of 1973.

Each of the preceding definitions shares common elements—diagnosis of mental illness, prolonged duration, and functional or role incapacity. Although there has been a developing consensus about the characteristics of persons who are psychiatrically disabled, there is by no means consensus on the precise operational definition of these characteristics (Bachrach, 1988a). The predominant characteristics of the persons who are the focus of this book reflect Goldman's description of a severe mental disorder resulting in disability of prolonged duration. These dimensions transcend traditional categories and describe a population characterized by significant vocational or social deficits and neuroticlike responses to sources of stress (Summers, 1981). Regardless of the particular diagnostic category, a population exists of people with mental illness who are simply not functioning well in their living, learning, and working environments (Adler, Drake, Berlant, Ellison, & Carson, 1987; Dion & Anthony, 1987; Pepper & Ryglewicz, 1988). It is these persons who benefit from the psychiatric rehabilitation approach.

Within this severely psychiatrically disabled population are subpopulations, such as young adults (e.g., Bachrach, 1982b; Harris & Bergman, 1987b; Pepper & Ryglewicz, 1984), persons who are homeless (e.g., Farr, 1984), senior citizens (e.g., Gaitz, 1984), persons with both a severe physical disability and severe psychiatric disability (e.g., Pelletier, Rogers, & Thurer, 1985), persons who are also developmentally disabled (e.g., Eaton & Menolascino, 1982; Reiss, 1987), and persons with substance abuse problems (e.g., Bachrach, in press; Foy, 1984; Talbot, 1986).

The psychiatric rehabilitation philosophy and technology described in this book are relevant to serving each of these subpopulations. Whether the subpopulations of persons with severe psychiatric disabilities are categorized by age (e.g., senior citizens, young adults), location (e.g., homeless) or additional diagnoses (e.g., physical disabilities, developmental disabilities, substance abuse), the psychiatric rehabilitation approach is a useful way to serve these subpopulations. The rehabilitation approach focuses on improving function and role performance and is not relegated to irrelevance due to factors such as age, location, or additional diagnoses.

Distinctions between Treatment and Rehabilitation

This section provides a summary of the distinctions between treatment and rehabilitation and highlights the unique contributions of the rehabilitation approach, illustrating how treatment and rehabilitation complement one another. The practice of psychiatric rehabilitation and the practice of psychiatric treatment (e.g., psychotherapy, chemotherapy) overlap to an extent. Psychiatric treatment and psychiatric rehabilitation procedures ideally occur in close sequence or simultaneously. Also, treatment techniques and rehabilitation techniques are often carried out in the same program or in separate programs run by the same agency. In addition, psychiatric treatment and psychiatric rehabilitation are sometimes provided by the same practitioner.

However, distinctions must be made between treatment and rehabilitation if the field of psychiatric rehabilitation is to contribute fully and effectively to the needs of persons with psychiatric disabilities. Maximal contributions by psychiatric rehabilitation practitioners cannot occur if the field of psychiatric rehabilitation continues to be perceived as the so-called poor sister of psychiatric treatment. All too frequently in the past psychiatric rehabilitation was either not considered at all, considered only after the treatment had concluded, or considered as an alternative when treatment had failed. Unfortunately at present, many psychiatric rehabilitation programs use treatment techniques rather than the technology of psychiatric rehabilitation. In many rehabilitation settings (e.g., group residences) the main helping modality is apt to be chemotherapy combined with individual or group therapy (sometimes euphemistically called group discussion). In addition, psychiatric rehabilitation practitioners are often overly concerned with psychiatric diagnosis, even though rehabilitation research indicates that the diagnostic label itself provides the staff with little information relevant to prescribing the rehabilitation intervention or predicting the rehabilitation outcome. Furthermore, the staff providing rehabilitation services are often selected for their training in the use of therapeutic techniques and not for their expertise in psychiatric rehabilitation technology.

Closer examination of rehabilitation programs often reveals

7

that the differences between treatment and rehabilitation do not extend to what actually occurs in practice. Only the positions of the players have changed; the rules remain the same. In fact, rehabilitation services are often evaluated in terms of the therapeutic model used. Evaluations are positively influenced if the rehabilitation program includes psychotherapy or has highly paid treatment consultants. *Treatment* continues to connote more status than does *rehabilitation*. It is not uncommon for professionals in a psychiatric rehabilitation setting to feel compelled to mention that in addition to their more mundane chores they also do therapy.

The rehabilitation professional, whether a psychiatrist, rehabilitation counselor, psychologist, social worker, nurse, occupational therapist, or mental health counselor, must be able to answer this question: "What can I effectively do for persons with psychiatric disabilities that is not some variation of chemotherapy or psychotherapy?" The psychiatric rehabilitation field must be defined by *what it is* rather than by what it is not. The unique philosophy and technology of psychiatric rehabilitation need to be adopted so that mental health professionals who practice psychiatric rehabilitation can be trained in a specific set of skills built on knowledge of the psychiatric rehabilitation process.

Table 1–2 provides an overview of the traditionally perceived differences between treatment and rehabilitation. The first column in Table 1–2 illustrates that the mission of rehabilitation is to improve functioning in specific environments; that rehabilitation interventions can occur without any underlying theory of causality; that the focus of the rehabilitation intervention is not on past causes but on current functioning and future goals; that the rehabilitation diagnosis focuses on a person's skills and environmental supports; that the rehabilitation intervention is based on this rehabilitation diagnosis and attempts to develop skills and increase supports; and that the historical roots of rehabilitation are human resource development, vocational rehabilitation, special education and learning approaches, physical medicine and physical rehabilitation, and client-centered therapy. The second column of Table 1–2 presents the emphases of the treatment approach.

In the past the basic difference between psychiatric treatment and psychiatric rehabilitation has been typically described in terms of philosophy. Rehabilitation was directed primarily at

TABLE 1–2 *Traditionally Perceived Differences*
between Rehabilitation and Treatment

	Rehabilitation	*Treatment*
Mission:	Improved functioning and satisfaction in specific environments	"Cure," symptom reduction, or the development of therapeutic insights
Underlying causal theory:	No causal theory	Based on a variety of causal theories that determine the nature of the intervention
Focus:	Present and future	Past, present, and future
Diagnostic content:	Assess present and needed skills and supports	Assess symptoms and possible causes
Primary techniques:	Skills teaching, skills programming, resource coordination, resource modification	Psychotherapy, chemotherapy
Historical roots:	Human resource development; vocational rehabilitation; physical rehabilitation; client-centered therapy; special education and learning approaches	Psychodynamic theory; physical medicine

developing the person's strengths, focusing on the person's assets as a way to restore his or her capacity to function in the community. In contrast, psychiatric treatment was typically directed at reducing the person's symptoms, focusing on liabilities in order to alleviate the person's symptoms.

Unfortunately, this philosophical distinction has not been translated into differences in practice. Psychiatric rehabilitation practice has remained for the most part similar to treatment practices. The training programs designed to teach psychiatric rehabilitation practitioners unique psychiatric rehabilitation skills are either nonexistent or primarily teach the theory and techniques of treatment. The philosophical distinction between treatment and rehabilitation has affected the practice of psychiatric rehabilitation only to the extent that the setting in which rehabilitation occurs may be different from the treatment setting.

This book defines the unique philosophy and technology of psychiatric rehabilitation. Although the complete process of helping persons with psychiatric disabilities includes both treatment and rehabilitation, they should be separated conceptually so that the rehabilitation process receives recognition for its own unique contribution to helping persons with psychiatric disabilities.

The Need for a Psychiatric Rehabilitation Approach

The need for a psychiatric rehabilitation approach to serve persons with severe psychiatric disabilities is well documented. As a matter of fact, it is probably the one approach that seems to be of interest to a wide range of persons and organizations, including family members, self-help and advocacy groups, NIMH, the Rehabilitation Services Administration, and many state mental health agencies. Furthermore, data exist that document the extensive rehabilitation needs of this group (Solomon, Gordon, & Davis, 1983; Wasylenki, Goering, Lancee, Fischer, & Freeman, 1981). Paradoxically, however, psychiatric rehabilitation has been of little interest to university educators of mental health professionals until recently (Anthony, Cohen, & Farkas, 1988; Bevilacqua, 1984; Talbot, 1984).

The National Alliance for the Mentally Ill (NAMI) has become the major voice of the relatives of persons who are severely psychiatrically disabled. A national survey of NAMI members conducted by Spaniol (Spaniol & Zipple, 1988) as well as other studies (Castaneda & Sommer, 1986) have reported that family members experienced the greatest need for improved social and vocational rehabilitation services. Similarly, a survey that included consumer advocates (Lecklitner & Greenberg, 1983) indicated that one strategy thought to have the most impact on persons with severe psychiatric disabilities was an "emphasis on rehabilitation approaches." (p. 428)

The need for improved rehabilitation services was dramatically illustrated by the aftermath of the deinstitutionalization movement (Farkas & Anthony, 1981). The two-thirds reduction in available state hospital beds resulted in a significant increase in the number of persons with severe psychiatric disabilities whose place of residence changed from the back ward to the back street to, in many cases, the main street.

With increasing visibility, the functional limitations of some of those persons quickly became apparent, and in 1977 NIMH launched the Community Support Program (CSP). This program was designed as a pilot federal and state collaboration to explore strategies for delivering community-based services to persons with severe psychiatric disabilities. National data on the persons initially served by CSP illustrate the extreme functional limitations of this group. For example, their median yearly income is $3,900; 50% receive Social Security benefits, approximately 10% are competitively employed, and of the unemployed only 9% are actively searching for work; only 12% are married; 71% rarely or never engage in recreational activities with others, indicating that the lack of work behavior has not been replaced by recreational activities (Goldstrom & Manderscheid, 1982). A more recent survey of CSP clients found a similar level of disability (Mulkern & Manderscheid, 1989).

The NAMI survey data compiled by Spaniol also attest to the functional incapacities of this population. Relatives report that about 5% of their family members with psychiatric disabilities are engaged in full-time competitive employment, even though 92% have a high school education and 60% have either post-high school training or attended college! It is no wonder, then, that families desire improved rehabilitation services (Spaniol & Zipple, 1988). Spivak, Siegel, Sklaver, Deuschle, and Garrett (1982) interviewed 99 long-term patients of a community mental health center. They reported that the group was "distinguished by low levels of educational, financial, and vocational achievement; only 13% were working more than half-time even though at intake about two-thirds were judged capable of work" (p. 241). These authors suggest that the data point to the need for a rehabilitation approach, focusing on the "accomplishment of tasks and the development of skills."

While the data on the function and role performance of persons with severe psychiatric disabilities indicate the need for rehabilitation, rehabilitation services are typically still not provided (Wasylenki et al., 1985). For example, a group of 550 patients discharged from two state hospitals were followed for one year

(Solomon et al., 1983). Most of these ex-patients received some type of case management, individual therapy, and chemotherapy. Only a small minority received rehabilitation services. "It is highly likely that many more patients than currently receive them could benefit from social and vocational services. A much higher proportion than those who received rehabilitation-oriented services was assessed as needing these services by the social workers at the time of discharge from the hospital" (Solomon et al., 1983, p. 39).

A Look at the Future

Client data and various surveys indicating the need for improved rehabilitation services stand in marked contrast to the dearth of effective psychiatric rehabilitation services and skilled psychiatric rehabilitation practitioners. The short-term growth of psychiatric rehabilitation is based on a documented need for these services. It is hoped that the long-term need for continued development of psychiatric rehabilitation services will be based on empirical studies of its effectiveness—studies undertaken as the field grows.

The stage has been set for a rehabilitative approach to emerge within mental health services. The unmet needs of those persons with psychiatric disabilities are many. Yet, if the field of psychiatric rehabilitation is going to make a meaningful contribution to the mental health service system and ultimately to those whom the mental health system is supposed to serve, it must be based on more than propitious timing and individual suffering. For these factors, motivating as they are, offer no guarantee of improved and effective services. The ultimate efficacy of the psychiatric rehabilitation field will be based on improving the attitudes, knowledge, and skills of those who practice it, the structure of the programs in which they practice, and the characteristics of the service systems that support their practice.

2

Review of
the Research:
Historical Myths

*A man should never be ashamed to own he has been
wrong, which is but saying in other words, that he is
wiser today than he was yesterday.*

Jonathan Swift

The times they are a-changing—and well they should.
But not every past development was irrelevant to the goals
of rehabilitation. In fact, although the psychiatric rehabilitation field is burdened by past errors, much has been learned.
This chapter discusses some of the historical developments in mental

---

Parts of this chapter are excerpted with permission from the following:

Anthony, W. A., & Liberman, R. P. (1986). The practice of psychiatric rehabilitation: Historical, conceptual, and research base. *Schizophrenia Bulletin*,
12, 542–559.

Anthony, W. A., Kennard, W. A., O'Brien, W. F., & Forbess, R. (1986).
Psychiatric Rehabilitation: Past myths and current realities. *Community
Mental Health Journal*, *22*, 249–264.

Farkas, M. D., Anthony, W. A., & Cohen, M. R. (1989). An overview of
psychiatric rehabilitation: The approach and its programs. In M. Farkas
& W. A. Anthony (Eds.), *Psychiatric rehabilitation programs: Putting
theory into practice*, pp 1–27. Baltimore: Johns Hopkins University Press.

health and rehabilitation, focusing on their positive influence on present psychiatric rehabilitation practice. Also, fifteen myths that have slowed the adoption of the psychiatric rehabilitation approach are presented and explored in light of the knowledge from recent research studies.

Before examining the past with respect to psychiatric rehabilitation, however, the term *psychiatric rehabilitation* must be defined. As mentioned in chapter 1, the mission of psychiatric rehabilitation is to assist persons with long-term psychiatric disabilities increase their functioning so that they are successful and satisfied in the environments of their choice with the least amount of ongoing professional intervention (Anthony et al., 1983). In essence, psychiatric rehabilitation helps people to function better and be satisfied in their chosen communities. The process by which this is accomplished includes developing the persons' skills and/or developing more supports in their environments; in other words, helping people to change and/or their living, learning, or working environments to change (Anthony & Margules, 1974).

Relevant Historical Developments

The idea of a psychiatric rehabilitation approach has its roots in the nineteenth century, even though its general acceptance as a legitimate and credible intervention has occurred only within the last several decades.

Several historical developments most relevant to the development of a psychiatric rehabilitation approach are 1) the moral therapy era, 2) the inclusion of persons with psychiatric disabilities into the state vocational rehabilitation program, 3) the development of community mental health centers, and 4) the psychosocial rehabilitation center movement.

Moral Therapy Era

Several ideas consistent with psychiatric rehabilitation surfaced in the 1800s during the so-called moral therapy era. The nineteenth-century moral therapists emphasized several treatment principles that became part of current psychiatric rehabilitation practice. Moral treatment stressed a comprehensive assessment of a person with a psychiatric disability, examining the person's work, play, and social activities. In psychiatric rehabilitation the person's vocational, avocational, and social areas of functioning are diagnosed and developed. Also consistent with moral treatment which recognized that structured activity can have therapeutic value beyond that provided by verbal therapy, the goal of the psychiatric rehabilitation approach is to have the person act differently, and the major psychiatric rehabilitation interventions use activity rather than simply verbal therapy as the means to reach that goal.

State Vocational Rehabilitation

The state vocational rehabilitation program was initially designed for the rehabilitation of persons with physical disabilities. The 1943 amendments to the Vocational Rehabilitation Act made persons with psychiatric disabilities eligible for financial support and vocational rehabilitation services. These amendments provided legitimacy to the idea of rehabilitating people with psychiatric disabilities and grounded the practice of psychiatric rehabilitation in improving vocational functioning.

Vocational rehabilitation interventions have now become an integral part of the history and development of the psychiatric rehabilitation field. Vocational activity is a central ingredient in the Fountain House model of psychiatric rehabilitation (Beard, Propst, & Malamud, 1982). Although the last several decades have seen the focus of rehabilitation shift from vocational functioning to a focus on a person's functioning in all environments (e.g., residential, educational, social), the availability of funds and services targeted to the vocational rehabilitation of persons with psychiatric disabilities has significantly aided the practice of psychiatric rehabilitation.

Community Mental Health Centers (CMHC)

Much has been written about the failure of CMHCs to provide the comprehensive services needed by persons with severe psychiatric disabilities (Bassuk & Gerson, 1978). Because of the deinstitutionalization movement, many more persons with long-term psychiatric disabilities are living in the community and these persons are not a high priority population for CMHCs. Yet, the

15

CMHC concept of treatment and support near the person's home and work settings is consistent with the practice of psychiatric rehabilitation. Also incorporated in psychiatric rehabilitation practice is the CMHC idea of immediately intervening in persons' crises without removing them from their environment for long periods of time—if at all. Thus, some of the innovations of the CMHC initiative are very much a part of the psychiatric rehabilitation approach.

Psychosocial Rehabilitation Centers

Grob (1983) and Rutman (1987) have traced the origins and current status of psychosocial rehabilitation centers. The early centers, such as Fountain House and Horizon House, were founded by groups of ex-patients for the purpose of mutual aid and support. These social clubs gave birth to the early comprehensive, multiservice psychosocial rehabilitation centers such as Thresholds in Chicago; the Social Center for Psychiatric Rehabilitation in Fairfax, Virginia; Center Club in Boston; Fellowship House in Miami; Hill House in Cleveland; and Portals House in Los Angeles. In these centers the philosophy of the psychiatric rehabilitation approach is put into practice. A fundamental tenet of the psychiatric rehabilitation approach, and a principle basic to the operation of these centers, is that rehabilitation is designed to improve the competencies of persons with psychiatric disabilities. From the beginning, these centers have emphasized strategies to help people cope with the environment rather than succumb to it (Wright, 1960) and have stressed health induction rather than symptom reduction (Leitner & Drasgow, 1972). Beard has said that a fundamental belief of these centers is that persons with severe psychiatric disabilities have the capacity to be productive (Beard et al., 1982).

The development of psychosocial rehabilitation centers has also served to focus the goals of the psychiatric rehabilitation approach on behavioral change in a person's environment of need. These centers have not valued the development of therapeutic insight achieved through verbal therapies (Dincin, 1981). Their orientation has been toward reality factors rather than intrapsychic factors, on improving a person's ability to perform in a specific environment, even in the presence of a residual disability (Grob, 1983).

Psychosocial rehabilitation centers have played a significant role in the development of the psychiatric rehabilitation approach. Their influence in the immediate future should be even greater. Since 1976, Fountain House has conducted the largest psychiatric rehabilitation training program in the world, initially under a grant from NIMH (Fountain House, 1976). The purpose of this grant was to assist CMHCs and other types of mental health settings in establishing rehabilitation services based on the psychosocial rehabilitation center model developed by Fountain House. In a 5-year period, individuals from agencies located in 38 states, the District of Columbia, Sweden, and Canada were trained. During this time the number of psychosocial rehabilitation settings has increased from 18 to 148. As Fountain House continues its training effort, and the number of psychosocial rehabilitation settings continues to grow, the presence of settings in which to conduct future research is ensured.

Gradual Recognition of the Differences between Treatment and Rehabilitation

As the field of psychiatric rehabilitation evolved, it became easier to distinguish the differences between psychiatric treatment and psychiatric rehabilitation. As highlighted in chapter 1, the fundamental difference between treatment and rehabilitation was typically described in terms of philosophical goals. Treatment focuses on decreasing a person's symptoms or pathology, while psychiatric rehabilitation focuses on developing a person's strengths or assets. Treatment alleviates a person's dysfunction; rehabilitation restores a person's function. Leitner and Drasgow (1972) characterized the rehabilitation philosophy as "health induction," in contrast to the treatment philosophy of "sickness reduction." They maintained that our past efforts at helping persons who are psychiatrically disabled have been directed more at minimizing sickness than maximizing health. In a similar vein Martin (1959) used the term rehabilitation to refer to those "activities which attempt to discover and develop a patient's assets in contrast

17

to treatment which is a direct attack on a patient's disability" (p. 56).

Chapter 1 outlines the most basic distinctions between treatment and rehabilitation that are now generally recognized. In contrast to the past, the psychiatric rehabilitation field is now defined by more than the absence of a traditional psychotherapeutic emphasis and the presence of an overall goal of health induction. Rehabilitation is becoming a complement to psychiatric treatment.

Dispelling the Myths of the Past

In order for the field of psychiatric rehabilitation to develop further, and to facilitate the adoption of its philosophy and technology by mental health settings, a number of myths have to be discarded. Not only have these myths retarded the field of psychiatric rehabilitation, they have also burdened the field of mental health in general. Research in the 1960s, 1970s, and 1980s proved these beliefs to be false. The fifteen myths listed in Table 2–1 must no longer be allowed to hinder future developments in psychiatric rehabilitation.

Fifteen Myths

Myth One: The majority of people with psychiatric disabilities are being successfully rehabilitated. In fact, the majority of persons who leave the hospital return for additional treatment, often repeatedly. Furthermore, only a small percentage of persons discharged from the hospital work in competitive employment.

Many studies provide estimates of the recidivism base rate for hospitalized psychiatric patients who receive the traditional hospital regimen of drug therapy and/or individual or group therapy. Although the studies differ in dates, samples, geographic location, and type of institution, their results are remarkably similar: they suggest a recidivism rate for a 1-year period of from 35–50%. Estimates of recidivism at 3–5 years reach 65–75% (reviewed by Anthony, Buell, Sharratt, & Althoff, 1972; Anthony, Cohen, & Vitalo, 1978). Reviews of the employment literature over the last 10 years have provided similarly discouraging data with respect

TABLE 2–1 *Fifteen Myths*

1. The majority of people with psychiatric disabilities are being successfully rehabilitated.

2. Increasing compliance with drug treatment can singularly affect rehabilitation outcome.

3. Traditional inpatient therapies, such as psychotherapy, group therapy, and drug therapy, positively affect rehabilitation outcome.

4. Innovative inpatient therapies, such as milieu therapy, token economies, and attitude therapy, positively affect rehabilitation outcome.

5. Hospital-based work therapy positively affects employment outcome.

6. Time-limited community-based treatment produces better rehabilitation outcome than does time-limited hospital-based treatment.

7. Community-based treatment settings are well used by persons who are psychiatrically disabled.

8. Where a person is treated is more important than how a person is treated.

9. Psychiatric symptomatology is highly correlated with future rehabilitation outcome.

10. A person's diagnostic label provides significant information relevant to a person's future rehabilitation outcome.

11. A strong correlation exists between a person's symptomatology and a person's skills.

12. A person's ability to function in one type of environment (e.g., a residential setting) is predictive of a person's ability to function in a different type of environment (e.g., a vocational setting).

13. Rehabilitation outcome can be accurately predicted by professionals.

14. A person's rehabilitation outcome is a function of the credentials of the mental health professional with whom the person interacts.

15. A positive relationship exists between rehabilitation outcome and the cost of the intervention.

to the employment outcome of formerly hospitalized psychiatric patients (Anthony et al., 1972, 1978; Anthony & Nemec, 1984). The earlier reviews of employment data suggested that during the year following discharge, approximately 20–30% of ex-patients either worked full-time throughout the year or were employed at the 1-year follow-up date. The later studies that focused on more severely disabled psychiatric patients suggest an employment rate of less than 15%, with some of the most recent follow-up studies reporting a 0% employment rate for long-term patients targeted for deinstitutionalization (Farkas, Rogers, & Thurer, 1987).

Variations of these two outcome criteria—recidivism and employment—have been used almost exclusively in early psychiatric rehabilitation outcome studies and have clear advantages. They are objective, have meaning to lay persons, translate readily into economic benefits, and make it possible to compare studies using similar outcome measures. In spite of these advantages, these two outcome criteria have numerous disadvantages. A broader range of outcome measures has been recommended for use in future studies (Anthony et al., 1972, 1978; Anthony & Farkas, 1982; Bachrach, 1976a; Erickson, 1975; Mosher & Keith, 1979) and are in fact currently being used (see chapter 3).

Myth Two: Increasing compliance with drug treatment can singularly affect rehabilitation outcome. The advent of drug therapy 3 decades ago was thought by some to eliminate the need for rehabilitation. Unbridled enthusiasm for drug therapy led many people to think that the ultimate answer had come:

> [These new drugs] were hailed as the solution to the problems of nearly all mental illness—they were the panacea, and to the more enthusiastic, they were the means by which most if not all previous forms of treatment could be eliminated and mental illnesses could be eradicated. The popular press was filled with dramatic examples of work done with the drugs. (Felix, 1967, p. 86)

Now it is realized that the goals achieved by drug treatment, while still dramatic, are somewhat more modest: reduction of certain psychiatric symptoms, reduced use of physical restraints, an increase in the time spent by clients in various forms of therapeutic activities, and an increase in psychiatric hospital discharges (Anthony, 1979). As pointed out in regard to myth one, the majority of persons with psychiatric disability are not being rehabilitated, even with drug therapy. Three decades after the discovery of the first antipsychotic drug, there is little evidence that drug therapy increases a client's strengths and assets. Chemotherapy alone cannot improve a person's ability to interview for a job, converse with friends, respond to another person's feelings, or program a computer (Englehardt & Rosen, 1976). Furthermore, a series of

20

studies conducted in the United States, Great Britain, and France has shown that increased drug treatment compliance does not significantly reduce a person's risk of relapse (Schooler & Severe, 1984).

Drug treatment and rehabilitation treatment are still viewed as complementary interventions. However, increased drug treatment compliance does not eliminate the need for psychiatric rehabilitation.

Myth Three: Traditional inpatient therapies, such as psychotherapy, group therapy, and drug therapy, positively affect rehabilitation outcome. This myth is related to myths one and two. It was believed that patients were being rehabilitated using drug therapy and traditional therapies. Yet now we know that these traditional inpatient treatment methods do not differentially affect rehabilitation outcome (first reviewed by Anthony et al., 1972). These methods were originally designed solely to reduce symptoms or to provide therapeutic insight, and little evidence exists suggesting they can do more than this.

Although some inpatient treatment techniques (e.g., token economies) have resulted in dramatic improvements in patients' behavior in the hospital, they have not demonstrated a similar effect on community functioning. In-hospital behavior simply does not correlate strongly with community behavior (Erickson, 1975; Erickson & Hyerstay, 1980). To a large extent the behavior of persons with psychiatric disabilities, similar to the behavior of nondisabled persons, is determined by the situation in which the behavior occurs; and the behaviors demanded by the hospital environment do not necessarily resemble behaviors needed for functioning in the community.

However, several innovative inpatient programs, all comprehensive in nature and connected to community programming, have demonstrated an effect on community functioning (Becker & Bayer, 1975; Carkhuff, 1974; Heap, Boblitt, Moore, & Hord, 1970; Jacobs & Trick, 1974; Paul & Lentz, 1977; Waldeck, Emerson, & Edelstein, 1979). These approaches are characterized by a primary focus on developing patient skills, creating an atmosphere in which the treatment staff believe themselves to be an important component of the treatment program, and outreach into the community.

Myth Four: Innovative inpatient therapies, such as milieu therapy, token economies, and attitude therapy, positively affect rehabilitation outcome. It was thought that if traditional inpatient therapies did not make a difference, perhaps an innovative treatment approach might. New total-push therapy procedures attempt to structure the total hospital environment so that most of the patients' waking hours are directed at therapeutic ends. Although these procedures (variously described as milieu therapy, attitude therapy, social learning therapy, and token economies) differ in terms of their theoretical base and the techniques used to facilitate change, they are similar in that they all therapeutically structure the patient's hospital environment.

From a rehabilitation perspective, these programs are similar in a more basic way. Although all these approaches have been able to demonstrate positive effects on within-hospital behavior, they have as yet not demonstrated their effects on measures of community adjustment. Outcome studies of these approaches have typically confined their analysis to changes in the patients' ward behavior (e.g., Foreyt & Felton, 1970).

These innovative treatments held great promise. To their credit, the proponents of these treatments researched their effectiveness on out-of-the-hospital functioning. Nevertheless, these treatments, conducted only on an inpatient basis, clearly cannot effect rehabilitation outcome.

Myth Five: Hospital-based work therapy positively affects employment outcome. Two decades ago Kunce (1970) surveyed the literature on work therapy and concluded that the research does not support the idea that inpatient work therapy can impact rehabilitation outcome. He estimated that regardless of whether patients receive work therapy, 33% will become employed, a finding consistent with the base-rate data presented earlier. One study (Walker, Winick, Frost, & Lieberman, 1969), which contrasted two types of work therapy, reported that at a 6-month follow-up 36% of both groups held regular competitive work at some time during the six months.

Some researchers (Barbee, Berry, & Micek, 1969) have suggested that work therapy may foster institutional dependency. Their results indicated that patients participating in work therapy remained longer in the hospital than did nonparticipants. During the 2-year follow-up, 46% of the work therapy group and 23% of the nonparticipants were readmitted to the same hospital. However, when readmission to all psychiatric facilities was examined, there were no significant differences between groups. More recent reviews of vocational rehabilitation research with the more severely psychiatrically disabled persons, found lower employment percentages but reported no studies attesting to the efficacy of hospital-based work therapy in impacting employment outcome (Anthony et al.,1978; Anthony, Howell, & Danley, 1984).

Myth Six: Time-limited community-based treatment produces better rehabilitation outcome than does time-limited hospital-based treatment. This is a tough myth to debunk. It has become relatively easy to admit that hospital programs aren't working, yet many people have believed that almost anything attempted in the hospital can be done better in the community. In general, however, this has not been the case.

The literature concerning the relative effectiveness of institutional versus alternative placement for persons with psychiatric disabilities is both sparse and somewhat contradictory. Test and Stein (1978) reviewed a series of studies (Davis, Dinitz, & Pasamanick, 1974; Herz, Spitzer, Gibbon, Greerspan, & Reibel, 1974; Langsley, Machotka, & Flomenhaft, 1971; Langsley & Kaplan, 1968; Michaux, Chelst, Foster, & Pruin, 1972; Mosher & Menn, 1978; Pasamanick, Scarpitti, & Dinitz, 1967; Polak, 1978; Rittenhouse, 1970; Stein & Test, 1978; Test & Stein, 1977, Wilder, Levin, & Zwerling, 1966) that compared various community alternatives to in-hospital treatment on the following outcome variables: time out of the hospital and readmission rates, psychiatric symptomatology, psychosocial functioning (e.g., role performance, employment, and social functioning), and client satisfaction. After discussing the methodological caveats concerning comparability of the design and quality of the outcome measures and noting the diversity of treatment modalities, duration of treatment, and methods for outcome measurement, Test and Stein (1978) observed that the results were similar for certain outcome measures. They concluded that community treatment initially results in less time spent in the hospital, but after 1 year the difference disappears; that there is no difference in the amount of symptom reduction between community alternatives and in-hospital treatment; and

23

that there is no difference between the in-hospital and community treatment groups in the amount of change in psychosocial functioning.

Dellario and Anthony (1981) reviewed additional studies not cited by Test and Stein (1978) and arrived at similar conclusions—*once treatment is withdrawn*, there is no significant difference between the two treatment settings on symptom reduction, psychosocial functioning, instrumental functioning, and personal adjustment. Initial differences that have been reported regarding rehospitalization, time spent in the hospital, and employment tend to wash out by 18 months *following treatment termination*. Although there are some exceptions, (for example, the Weinman, Kleiner, Yu, & Tillson, 1974, study indicated significant differences in rehospitalization at 24 months), the weight of evidence seems to support the conclusion that, regardless of the type of community-based alternative, there is no long-lasting superiority of time-limited, community-based alternative treatment compared to time-limited, in-hospital treatment. It also is evident that, without interventions designed to affect sustained long-term outcome directly, sustained long-term progress on any outcome indicator should not be expected, regardless of the institutional or community location of the treatment setting.

Interestingly, both reviews (Dellario & Anthony, 1981; Test & Stein, 1978) did report significant differences in client satisfaction, favoring the community settings. Consumers remind us that this is an important outcome variable. However, these differences in satisfaction may not reflect program differences but rather the fact that community programs offer the client relatively more freedom.

The most frequently cited review of comparative studies of institutional versus community-based treatment was authored by Kiesler (1982), who reviewed 10 studies in which persons with severe psychiatric disabilities were randomly assigned to either inpatient care or some alternative mode of outpatient care. Kiesler concluded that in no case were the outcomes of hospitalization more positive than alternative community treatment. However, in contrast to the other two reviews, Kiesler did not report the community program's effect once the program ended (i.e., *time-limited* community treatment). The implication of the need for long-term as opposed to time-limited community care is obvious. It has now become an accepted principle of psychiatric rehabilitation that persons with psychiatric disabilities should have the opportunity for long-term rehabilitation interventions if that is what they want and need.

Myth Seven: Community-based treatment settings are well used by persons who are psychiatrically disabled. Although persons who use the community services may be more satisfied with them than with hospital services, the overall utilization figures do not attest to their popularity. At issue is the inability of the mental health system to convince many persons with psychiatric disabilities both to accept and also to remain in community-based treatment. Wolkon (1970) has reported one study in which approximately two-thirds of patients referred to an outpatient setting failed to appear for treatment. The drop-out rate for state-of-the-art psychotherapy and medication management has been as high as 42% by 6 months, 56% by 1 year, and 69% for 2 years (Stanton et al., 1984). Attrition rates for psychosocial rehabilitation centers vary depending on the length of follow-up. Bond, Dincin, Setze, & Witheridge (1984) found a 3-month drop-out rate of more than 25% and reported that, even with considerable self-selection, drop-out rates exceeded 50% during the first 9 months. Data from Wolkon and Tanaka (1966) indicated a discontinuance rate at a psychosocial rehabilitation center of 60% during a 1-year period. Equally discouraging are the statistics provided by Sue, McKinney, and Allen (1976), who reported that of the 13,450 clients seen in 19 mental health facilities, 40% terminated treatment after one session. Community clinics must determine not only the type of patient who can most benefit from their services, but also how to ensure that this type of patient actually does appear and continue in treatment.

Because of the inadequate use of aftercare clinics, the introduction of such services may not always produce the expected reduction in communitywide outcome figures. McNees, Hannah, Schnelle, & Bratton (1977) attempted to determine how recidivism has been affected by the development of aftercare programs in three Tennessee counties. Although countywide statistics revealed no clear reduction in recidivism rates, recidivism rates were substantially lower for individuals who contacted the aftercare program

25

than for those who did not contact the program. Program completion also appears to correlate with rehabilitation outcome.

Myth Eight: Where a person is treated is more important than how a person is treated. The mythical nature of this statement flows from the preceding myths. To ask about the relative effectiveness of institutions versus community-based alternatives implies that *where* services are delivered is the primary factor that determines the effectiveness of psychiatric rehabilitation. However, the body of evidence clearly demonstrates that *following the termination of treatment* in another setting, there is little appreciable difference between clients treated in institutions and those treated in community-based alternatives. This finding raises the possibility that the *where* of service delivery is not more important than the *what* and *how* of service delivery.

The issue of *where* versus *how* has often been cast as a hospital versus community argument. Taken to the extremes, this argument proposes either a return to institutionalization, or by contrast, the closing of all psychiatric hospitals.

Even though rehabilitation programs begun in the hospital and linked to the community can have a positive impact on rehabilitation outcome, proponents of psychiatric rehabilitation have always stated that the community is the preferred site for rehabilitation interventions—for the very real reason that persons with long-term mental illness actually want to and do live in the community (Center for Psychiatric Rehabilitation, 1989). However, many community programs are as suspect in their community support and rehabilitation orientation as the hospital wards they have replaced. Research data indicate that many community programs, like many hospital programs, lack the necessary rehabilitation mission statements, record keeping systems, intervention strategies, and specially trained staff (Farkas, Cohen, & Nemec, 1988).

The question of the relative effectiveness of institutional and community-based alternatives can be asked at two levels: their effectiveness relative to each other, and their effectiveness relative to their potential functions. The second question may be the more sensible one: hospital- and community-based approaches should be viewed neither as mutually exclusive nor as necessarily antagonistic, but rather as complementary sources of potential impact on a variety of criteria pertaining to psychiatric rehabilitation outcomes (Dellario & Anthony, 1981).

For example, Waldeck, Emerson, and Edelstein (1979) described a program designed to help psychiatrically disabled individuals make the transition from institutional living to community living. The program emphasized in-hospital training and systematic aftercare as two points on a continuum of treatment. The community orientation and evaluation program (COPE) consisted of a sequenced process of in-hospital skill training and community integration whereby independent functioning in the community is predicated on the development of successive and cumulative skill proficiencies that are necessary for the client to meet basic needs. The initial outcome data from this project indicated that during the first 9 months of operation, the program had an 86% rate of nonreturn. This program views in-hospital and community-based treatment modalities as complementary services rather than as mutually exclusive alternatives.

Myth Nine: Psychiatric symptomatology is highly correlated with future rehabilitation outcome. A number of studies illustrate the lack of relationship between a variety of assessments of psychiatric symptomatology and future ability to live and work independently (Ellsworth, Foster, Childers, Arthur, & Kroeker, 1968; Gaebel & Pietzcker, 1987; Green, Miskimins, & Keil, 1968; Gurel & Lorei, 1972; Möller, von Zerssen, Werner-Eilert, & Wuschenr-Stockheim, 1982; Schwartz, Myers, & Astrachan, 1975; Strauss & Carpenter, 1972, 1974; Wilson, Berry, & Miskimins, 1969). Although occasional studies do report a relationship between a type of symptom and rehabilitation outcome (McGlashen, 1987), the evidence is overwhelming that little or no relationship exists. For example, no symptoms or symptom patterns categorized by *DSM III* have been consistently and routinely related to individual work performance.

On occasion, the studies have generated data contrary to what might be expected. For example, Wilson et al. (1969) found future vocational performance to be related to higher levels of aggressiveness and depression. An analysis of all of these types of studies indicates that vocational performance does not correlate with tension, distress/alienation, antisocial behavior (Lorei, 1967);

depression, anxiety, paranoid hostility, deteriorated thought (Ellsworth et al., 1968); alertness, orientation, use of defenses (Green et al., 1968); anxiety, verbal hostility, depression (Gurel & Lorei, 1972); thought disorder, depression, flattened emotion (Strauss & Carpenter, 1974); confusion, mania, depression (Schwartz et al., 1975); and global psychopathological state (Möller et al., 1982). As Turner (1977) concludes, "the capacity to perform the work role cannot be wholly or even largely accounted for by the presence or degree of manifest pathology" (p. 39).

Current studies conducted by the Center for Psychiatric Rehabilitation have attempted to clarify the relationship between employment outcome and psychiatric symptomatology. This lack of any relationship has puzzled many psychiatric rehabilitation practitioners because intuitively they have sensed a relationship. In contrast to the general literature, the Center for Psychiatric Rehabilitation studied only persons who had a vocational goal and were actively participating in a psychosocial rehabilitation program (Center for Psychiatric Rehabilitation, 1989). This particular study suggested that when the sample is restricted only to persons actively involved in a vocational rehabilitation program, there is a modest relationship between psychiatric symptomatology and employment outcome. However, the conclusions of Turner (1977) as reflected in his previously cited quote still hold.

Myth Ten: A person's diagnostic label provides significant information relevant to a person's future rehabilitation outcome. This myth flows directly from myth nine. Because of previous findings of no consistent relationship between psychiatric symptoms and rehabilitation outcome, we should expect no relationship between diagnosis and future independent living and vocational functioning. An overwhelming number of studies have confirmed the absence of such a relationship (Distefano & Pryer, 1970; Douzinas & Carpenter, 1981; Ethridge, 1968; Freeman & Simmons, 1963; Goss & Pate, 1967; Hall, Smith, & Shimkunas, 1966; Holcomb & Ahr, 1986; Lorei, 1967; Möller et al., 1982; Pietzcker & Gaebel, 1987; Sturm & Lipton, 1967; Watts & Bennett, 1977; Wessler & Iven, 1970). Even the belief that persons with bipolar affective disorder have drastically better rehabilitation outcomes has been questioned (Dion, Cohen, Anthony, & Waternaux, 1988). Rehabilitation outcomes for persons with multiple hospital admissions due to bipolar affective disorder approximate the rehabilitation outcomes of persons with multiple hospitalizations due to schizophrenia. The long-term nature of the illness rather than the specific symptoms seems to be the common denominator impacting rehabilitation outcome.

From a rehabilitation perspective, the third edition of the *Diagnostic and Statistical Manual for Mental Disorders* and its subsequent revision (American Psychiatric Association, 1980, 1987) are only small improvements over their predecessors. The inclusion of a global assessment scale (Highest Level of Adaptive Functioning Past Year, Axis V) and of a global assessment of the client's environment (Severity of Psychosocial Stressors, Axis IV) is to be commended, but their inclusion as optional categories "for use in special clinical and research settings" is unfortunate. No information on client skill functioning is included, nor is any method of specifying the environment in which the client needs or wants to function. As described in the manual's introduction, this third edition attempts to be more descriptive and can be used in the initial steps of treatment planning; however, it focuses on collecting information that has limited predictive value for psychiatric rehabilitation practitioners. As a document to be used in rehabilitating persons with long-term psychiatric disabilities, the *DSM III* leaves much to be desired.

The data attesting to the lack of relationship between the diagnostic label and rehabilitation outcome are voluminous. As the reviewed research suggests, the psychiatric diagnosis does not provide any unique, relevant information about the person's rehabilitation potential. This finding is really not that surprising as the psychiatric diagnostic system was developed to categorize symptom patterns, not to provide information about the rehabilitation prospects of persons with psychiatric disabilities.

Also problematic is the practice of psychiatric labeling, as presently conducted. The critical problems most often revolve around whether the psychiatric diagnostic system is reliable and valid, or whether the label may do more harm than good to the person. However, independent of the controversy surrounding the whole issue of psychiatric labeling, rehabilitation research indicates that the psychiatric diagnostic system has little to offer the rehabilitation approach.

This does not mean, however, that rehabilitation practitioners cannot use the information collected by traditional psychiatric diagnosticians in their attempts to diagnose persons with psychiatric disabilities. Rehabilitation practitioners can educate the traditional diagnosticians as to the information they need about rehabilitation outcome. Specifically, rehabilitation practitioners should ask the traditional diagnostician for any information they might have uncovered with respect to the person's goals, skills, skill deficits, interests, and interactions with significant others. Some of this information can be collected during a psychiatric diagnostic interview. The rehabilitation practitioner must be assertive in asking for this information and not settle merely for information about symptoms and labels (Anthony, 1979).

Myth Eleven: A correlation exists between a person's symptomatology and a person's skills. A major emphasis in rehabilitation practice is the assessment of a person's skills. This is a completely different procedure from diagnosing a person's symptoms because measures of skills and measures of symptoms show little relationship to one another. This is most apparent in studies that targeted either skills or symptoms as their treatment focus but took measures of both. For example, it is well known that hospitalization and drug treatment affect symptomatology yet have little impact on a person's vocational skills (Anthony et al., 1978; Ellsworth et al., 1968, Englehardt & Rosen, 1976). In particular, the research of Ellsworth and associates (1968) has shown that hospital treatment results in significant symptom reduction, but not changes in instrumental role performance. Examining this issue from a different perspective, Arthur, Ellsworth, and Kroeker (1968) found symptomatology, and not instrumental behavior, to be related to hospital readmission. Englehardt and Rosen (1976) concluded from their review of drug treatment that while chemotherapy impacts on symptomatology, "evidence for a direct effect of pharmacotherapy on the work performance of schizophrenic patients is so far lacking" (p. 459). Dion et al. (1988) reported that at 6 months after hospital discharge 80% of their sample of persons with bipolar affective disorders were symptom free or mildly symptomatic, yet only 20% were working at their expected level of employment. Turner (1977) summarizes the results of his data in this way: "they substantially contradict the assumption that work failure or work performance is primarily or even largely a consequence of the presence or absence of symptoms'' (p. 36).

Thus, knowledge of a person's psychopathology does not provide much evidence of a person's functional capacity. A rehabilitation diagnosis with its focus on skills and supports relevant to the person's goals must precede a rehabilitation intervention— just as a psychiatric diagnosis precedes psychiatric treatment.

There are suggestions that future research into different ways of categorizing symptoms (e.g., positive and negative symptoms) might show a correlation between symptoms and skills (Dion & Dellario, 1988). However, this will not obviate the need for a rehabilitation diagnosis to prescribe the rehabilitation intervention.

Myth Twelve: A person's ability to function in one type of environment (e.g., a residential setting) is predictive of a person's ability to function in a different type of environment (e.g., a vocational setting). Various estimates of persons' functioning in their community environment have been used by researchers. These estimates include global ratings of social adjustment, ratings of community adjustment, and measures of recidivism. On the basis of several decades of study, the conclusion of researchers is that functioning in one area shows little or no relationship to functioning in other areas (Avison & Speechley, 1987). It is now standard practice in outcome research to assume little or no relationship between measures taken in two different areas of functioning (Anthony & Farkas, 1982; Schwartz et al., 1975) and that the best predictor of outcome in a particular area of functioning is a previous measure of that area of functioning (Anthony, 1979; Möller et al., 1982).

The large scale research of Ellsworth and associates (1968) showed that the situation itself is a powerful determinant of a psychiatrically disabled person's ability to function. Their research found no relationship between hospital-based ratings of adjustment and community-based ratings of adjustment (Ellsworth et al., 1968). Forsyth and Fairweather (1961) had previously reported similar findings with respect to the independence of hospital and community adjustment measures.

A number of researchers have reported only a slight relationship (Forsyth & Fairweather, 1961; Freeman & Simmons, 1963; Gregory & Downie, 1968; Lorei & Gurel, 1973) or no relationship

(Arthur et al., 1968; Wessler & Iven, 1978) between recidivism and posthospital employment. The lack of a strong relationship between recidivism and employment is somewhat surprising in that those individuals who are working throughout the follow-up period cannot by definition be recidivists. Nevertheless, a significant percentage of persons appear to work and still recidivate, and other persons aren't able to work yet still do not become recidivists.

Other researchers have noted the independence of measures of vocational functioning from other types of assessments (Gaebel & Pietzcker, 1987). In Tessler and Manderscheid's (1982) study of more than 1,400 clients with severe psychiatric disabilities, very low correlations between remunerative employment and social activity (.11) and basic living skills (.16) were reported. They concluded that the "results supported the view that community adjustment involves relatively distinct yet independent dimensions" (Tessler & Manderscheid, 1982, p. 206). In several studies of state vocational rehabilitation clients, (a portion of which were psychiatrically disabled) the independence of the vocational dimension from other measures of adjustment was again confirmed. The results of Bolton (1974), replicated by Bolton (1978) and Growick (1979), showed that measures of vocational success are unrelated to self-reported changes in psychological adjustment.

Strauss and Carpenter (1972, 1974) examined predictors of outcome and focused specifically on four dimensions of outcome: work, symptoms, social relationships, and need for hospitalization. These investigators concluded that each of these areas must be considered separately in the evaluation of the functioning of persons with psychiatric disabilities. Examining the concept of situational specificity from still another perspective, several studies (Lorei & Gurel, 1973; Walker & McCourt, 1965) have found no correlation between work activity in the hospital and subsequent employment. Walker and McCourt (1965) reported that only 26% of the patients participating in work activity in the hospital were employed after discharge; furthermore, 20% of the patients who did not participate in work activity in the hospital were employed at follow-up.

In summary, the research clearly indicates that a person's

vocational capacity cannot be inferred from that person's daily nonvocational functioning, and vice versa.

Myth Thirteen: Rehabilitation outcome can be accurately predicted by professionals. Although professionals may describe various aspects of a person's behavior on rating forms that statistically correlate with posthospital employment, no data suggest that they know how to use these ratings to make an accurate prediction (Miles, 1967). That is, professionals do not know on which behavior to focus when making their clinical predictions.

If the goal is to increase predictive accuracy, long-term prediction (3–5 years) could be improved by simply predicting that every psychiatrically disabled person will become a recidivist or unemployed! Using the base-rate figures presented in the discussion of myth two, these predictions would be correct approximately 75% of the time or more. Obviously, the psychiatric rehabilitation practitioner's diagnostic system must become more refined than predicting failure for everyone! A tentative direction for the refinement of predictive skills would be a sharper focus on indices of present skills in relationship to the level of skills expected either by significant others or by persons with psychiatric disabilities themselves (Anthony, 1979).

Thus, consumers must be wary when they hear professionals discussing a person's chances for rehabilitation. Are these professionals assuming they can predict behavior across situations (myth twelve), or are they assuming knowledge of functioning based on knowledge of symptom (myth eleven)? In either case they are dealing in myths rather than facts.

Myth Fourteen: A person's rehabilitation outcome is a function of the credentials of the mental health professional with whom the persons interacts. Professionals from a wide variety of disciplines (e.g., nursing, psychiatry, social work, occupational therapy, rehabilitation counseling, psychology, recreation therapy) are involved in the practice of psychiatric rehabilitation. The clients' rehabilitation outcome is not a function of these credentials (Anthony & Carkhuff, 1976). Nurses do no better than social workers, who do no better than psychologists, and so on. While certain programs must hire particular professionals for legal reasons (most often revolving around medication and psychological testing), it makes no empirical sense to designate personnel positions in programs by credentials (i.e., so many OT's, psychologists, rehabilitation counselors). Personnel must be assigned to programs based on their skills related to achieving positive outcomes. One cannot assume a relationship between a practitioner's skills and a practitioner's credentials. Just as a diagnostic label tells us little about a client's skills, a professional degree (i.e., RN, PhD, CRC, MD) tells us little about a practitioner's skills.

The skills needed to assist persons with severe psychiatric disabilities through the rehabilitation process are not the exclusive province of any one profession—nor for that matter the exclusive province of professionals. Carkhuff (1971) coined the term *functional professionals* to identify those individuals, who heretofore have been called nonprofessionals, paraprofessionals, companions, volunteers, lay professionals, and subprofessionals. Groups of individuals who have been referred to by these labels include college students, psychiatric aides, community workers, consumers, parents, and mental health technicians. Thus, the "functional professional" in the mental health field can be defined as a person who, lacking formal credentials, performs those functions usually reserved for credentialed mental health professionals. In psychiatric rehabilitation these functions include skills teaching, skills programming, resource coordination, and personal support.

Myth Fifteen: A positive relationship exists between rehabilitation outcome and the cost of the intervention. If a setting has highly paid staff and expensive facilities, it does not necessarily produce better rehabilitation outcomes. Cost and outcome are not significantly correlated (Dickey, Cannon, McGuire, & Gudeman, 1986; Gorin, 1986; Grinspoon, Ewalt, & Shader, 1972; Walker, 1972).

A variable contributing to the lack of relationship between cost and outcome is the fact that combining chemotherapy with expensive forms of psychotherapy does not produce more symptom improvement than can be achieved by chemotherapy alone. Neither type of treatment has been shown to have much effect on functioning. When psychotherapy is given to long-term patients, it may increase the cost of treatment but not necessarily the outcome (Grinspoon, et al., 1972). For example, a study of the social disability of patients 1 month after discharge from a number of different hospitals found no relationship between hospital per diem costs and the patient's social adjustment. As a matter of fact, ex-patients of the hospital with the highest per diem rate were less socially adjusted than were ex-patients from the hospital with the lowest per diem rate (Walker, 1972).

It would seem that, based on what can be learned from history, proponents of a rehabilitation approach within the mental health system should not cite reduced costs as a reason for adopting a rehabilitation approach. A comprehensive well-run rehabilitation approach should produce additional client benefits, but it might also produce additional or different types of costs.

A Current Myth?

With respect to schizophrenia, *DSM III* states, "The most common outcome is one of acute exacerbations with increasing residual impairments between episodes." (American Psychiatric Association, 1980, p. 185). Throughout most of the 1980s, and officially until the appearance of *DSM III-R,* the expectation still was that over the long-term schizophrenia was a deteriorative disease.

Perhaps this assumption of increasing deterioration *may not be a function of the disorder per se but rather of how the person with the disorder is treated by the environment* (Harding, Zubin, & Strauss, 1987). Environmental factors that could contribute to chronicity include institutionalization, adoption of a patient role, lack of skills and supports, reduced economic opportunities, lowered social status, medication effects, staff expectations, and loss of hope. Many of these environmental factors can be impacted by a rehabilitation approach and explain why a psychiatric rehabilitation intervention can impact chronicity.

Harding, Zubin, and Strauss (1987) cite a number of long-term research studies, including their own (Harding, Brooks, Ashekaga, Strauss, & Breier, 1987a; 1987b), suggesting that a deteriorating course is not the rule. "The possible causes of chronicity may be viewed as having less to do with the disorder and more to do with a myriad of environmental and other social factors interacting with the person and the illness" (Harding, Zubin, &

Strauss, p. 483). It is interesting to note that not so long ago persons with mental retardation were expected to live their lives in institutions. Now this is the exception rather than the norm. Has the very nature of the disorder changed? Or, has the way the environment (both the treatment and societal environment) changed? It would seem that similar gains in battling chronicity in persons with severe psychiatric disability would be made with an intensive effort at personal rehabilitation and societal rehabilitation.

Summary

As the psychiatric rehabilitation field moves into the future, a number of myths can be discarded—myths that have retarded the field's development. Not surprisingly, most of the research showing these 15 statements to be myths was conducted 10 to 20 years ago. The lag time between the accumulation of new knowledge in mental health and its application to practice is usually considerable.

By throwing overboard this mythical baggage we can set a course toward future innovation, as there is still so much we need to learn about psychiatric rehabilitation. By operating on wrong assumptions in the past, we acted as if we knew a lot more than we actually did! For example, we acted as if the location was more important than the intervention, that drug treatment could singularly affect rehabilitation outcome, that practitioners' credentials were more important than their skills, that a psychiatric diagnosis was a useful and prescriptive tool for rehabilitation. Free from these myths, rehabilitation practitioners can focus on their mission—helping persons who have experienced psychiatric disabilities to function successfully and be satisfied in the living, learning, working, and social environments of their choice, with the least amount of ongoing intervention by the helping professions.

As the next chapter suggests, the time has come for psychiatric rehabilitation to take its place as a recognized, credible approach, integrated with existing treatment modalities. Psychiatric rehabilitation is complementary to current treatments and should be viewed as an addition to mental health practitioners' repertoire.

3

Review of the Research: Current Realities

Science is simply common sense at its best—that is, rigidly accurate on observation, and merciless to fallacy in logic.

T. H. Huxley

Although early research helped dispel the myths of psychiatric rehabilitation, more recent research has helped define the current realities and future promise of psychiatric rehabilitation. The literature reviewed in this chapter focuses on the target population, outcomes, and types of interventions that characterize the field of psychiatric rehabilitation (e.g., Beard, Propst, & Malamud, 1982; Dincin, 1981; Grob, 1983; Lamb, 1982).

Parts of this chapter are excerpted with permission from:

Farkas, M., & Anthony, W. A. (1987). Outcome analysis in psychiatric rehabilitation. In M. J. Fuhrer (Ed.), *Rehabilitation outcome: Analysis and measurement* (pp. 43–56). Baltimore: Paul Brookes.

Dion, G., & Anthony, W. A. (1987). Research in psychiatric rehabilitation: A review of experimental and quasi-experimental studies. *Rehabilitation Counseling Bulletin, 30,* 177–203.

Studies Included in the Review

As mentioned in chapter 1, the *target population* of psychiatric rehabilitation is those persons who have become disabled due to severe psychiatric illness. There are several definitions of severe psychiatric disability. Studies included in this chapter are focused on persons whose characteristics were consistent with the definition currently used by the National Institute of Mental Health's Community Support Program (CSP), the Rehabilitation Services Administration's (RSA), and Goldman's (Goldman et al., 1981) definition of the "chronically mentally ill" (discussed in chapter 1).

The *outcomes* of the studies reviewed in this chapter are measures of either client behavioral change, or client and/or societal benefits (Anthony, 1984; Anthony & Farkas, 1982). Measures of client behavioral change provide an indicator as to whether the clients are able to act differently as a result of their involvement in a psychiatric rehabilitation intervention—that is, what skills (e.g., expressing anger) are they now performing, either in the rehabilitation setting (e.g., responding to criticism from supervisors at the sheltered workshop) or in a nonrehabilitation environment (e.g., conversing with parents at the dinner table). Measures of client and societal benefits are an indicator of the gains for the client and/or society as a result of psychiatric rehabilitation. Simple measures of recidivism and days spent in the hospital or community can be used as rough estimates of client and societal benefits. Other types of benefit measures are estimates of the clients' satisfaction with their life situations and measures of residential, educational, and vocational status (e.g., level of employment, degree of independent living, educational achievement).

The *interventions* used in these studies emphasize changing the clients' skills, changing the clients' environmental supports, or both. The assumption of psychiatric rehabilitation is that by changing the person's skills and/or environmental supports, benefits will accrue to the client and society. Studies involving skill development and support development interventions are reviewed, whether the author specifically called the intervention a psychiatric rehabilitation intervention or not.

In summary, this chapter reviews studies that meet the

inclusion criteria of focusing on the identified client population, outcome measures, and intervention techniques. That is, the population studied was comprised of persons with severe psychiatric disabilities; the outcomes focused on improved client performance of skills, client benefits, and/or societal benefits; and the intervention techniques were designed to develop client skills and/or environmental supports.

Major Intervention Studies Not Included in the Review

Several types of studies of skill development and support development interventions are not reviewed here, in particular, studies of (1) social skills training; (2) family management; and (3) the community support system initiative. These areas are excluded because they represent a somewhat discrete focus of inquiry in which periodic reviews have been published. The findings from these reviews have already been incorporated into the general psychiatric rehabilitation literature (Anthony, 1979; Anthony, Cohen, & Cohen, 1984).

Social Skills Training

The literature abounds with conceptual and experimental studies of social skills training (SST) with psychiatric clients. SST attempts to identify interpersonal dysfunction in the client, identify specific social skills deficits, examine the circumstances in which the dysfunction occurs, and identify the specific nature of the skill deficits. SST uses highly structured educational procedures to address these deficits. The intervention focuses on small units of behavior and emphasizes a gradual approximation to the ideal behavior by extensive rehearsal (Morrison & Bellack, 1984). Reviews of issues related to social skills training appear periodically in the literature (Brady, 1984; Hersen & Bellack, 1976; Liberman et al., 1986; Morrison & Bellack, 1984; Wallace et al., 1980).

In general, most studies of social skills training demonstrate effectiveness in reducing targeted social skill deficits in psychiatric patients. However, the efficacy of these interventions has usually been measured in pre-post and follow-up role-play assessments

39

(Morrison & Bellack, 1984). As Hersen and Bellack (1976) noted, the evaluation of the generalization of social skills training to natural environments has been mixed. Also, the durability of social skills training has been questioned, although a number of studies do report that the skills are often maintained at the time of follow-up (Monti et al., 1979). Perhaps the most important observation that Morrison and Bellack (1984) have made is that social skills training, in and of itself, has not made major differences in patients' lives. The studies have focused primarily on target behaviors and topographic outcome (Wallace et al., 1980). They argue that social skills training may be of best use in a "multi-component, skills-oriented training program." In conclusion, it appears that SST holds most promise as a specifically targeted component of a comprehensive program. The methodology of SST can be integrated with the skills teaching component of a psychiatric rehabilitation intervention. (See the further discussion of a skills development intervention in chapter 7.)

Family Management

Like social skills training, family management models (often called family psychoeducational approaches) have received a great deal of attention in the conceptual and experimental literature. The concept of targeting interventions to families exhibiting high levels of "expressed emotion" has been studied repeatedly (Falloon, Boyd, McGill, Strang, & Moss, 1982; Hogarty et al., 1986; Leff, Kuipers, Berkowitz, Eberbein-Vries, & Sturgeon, 1982; Vaughn & Leff, 1976). In general, these studies use interventions that involve structured, behavioral, problem-solving methods to deal with family issues. Most studies demonstrate the effectiveness of this approach in preventing major symptom recurrences and decreasing relapse rates. An ongoing multisite family management study, funded by NIMH, has been designed to investigate the effect of family management interventions on rate of relapse, need for lower levels of neuroleptic medication, and better social functioning (Schooler & Keith, 1983; Schooler, Keith, Severe, & Matthews, in press). It is hypothesized that the family management intervention will allow a reduction in the medication level, resulting in fewer side effects such as tardive dyskinesia and akathisia. It is also hypothesized that patients treated with family management intervention will be less troubled by depression and social withdrawal and will have fewer episodes of symptom exacerbation.

Historically, rehabilitation philosophy has emphasized including the family in a comprehensive rehabilitation approach (Power & Dell Orto, 1980). Rehabilitation practitioners who work with persons with physical disabilities or developmental disabilities use the family as a valuable resource in rehabilitation. In contrast, mental health practitioners who work with persons with psychiatric disabilities have been slower in perceiving the family as a resource (Spaniol, Zipple, & Fitzgerald, 1984). The development of structured family interventions has expanded the possibilities for involving the family as a resource in the rehabilitation of persons with psychiatric disabilities. (See the further discussion of family issues and rehabilitation in chapter 8.)

Community Support System

Another recent innovation related to the field of psychiatric rehabilitation is the federally funded Community Support System initiative (CSS). The CSS literature has been extensively reviewed, and the outcomes have been described (Test & Stein, 1978; Test, 1984; Braun et al., 1981; Bachrach, 1982a) and replicated (Hoult, 1986). The concept of a CSS was developed as a response to the multiple needs of long-term psychiatric clients in the community. The functions of a CSS include identifying clients in need of services; helping them meet basic needs; providing mental health care, crisis intervention, comprehensive psychosocial rehabilitation services, case management, housing supports, family supports, and advocacy services; and facilitating use of formal and informal helping systems (Stroul, 1989; Test, 1984).

Test (1984) reviewed the literature relevant to CSS and reported that:

1. The more comprehensive a CSS, the more effective it is in forestalling relapse;
2. A CSS that provides intensive, direct assistance and teaching in psychosocial areas is more likely to lead to significant changes in social adjustment;
3. Those CSSs that provide comprehensive, intensive assis-

tance and teaching are most promising for favorable outcomes but may hasten relapse for some patients.

The target population of both CSS and the psychiatric rehabilitation approach are persons with severe psychiatric disabilities; CSS and psychiatric rehabilitation share a rehabilitation philosophy and a set of principles basic to helping persons who are severely disabled. The description of a CSS, with its focus on system functions, defines the context in which the psychiatric rehabilitation approach should be provided. Furthermore, ingredients of a psychiatric rehabilitation intervention are seen as major elements in a number of the components of a CSS (Mosher, 1986). The technology of psychiatric rehabilitation, especially its intervention strategies, has implemented the CSS philosophy and functions (Anthony & Stroul, 1986). In summary, the CSS initiative and the psychiatric rehabilitation approach are compatible with respect to their target population, their philosophy, service delivery context, and intervention strategies. (See the further discussion of CSS in chapter 10.)

Residential, Educational, and Vocational Status of Persons with Severe Psychiatric Disabilities

Before reviewing the impact of specific psychiatric rehabilitation interventions on outcomes for persons with long-term psychiatric disabilities, the typical residential, educational, and vocational status of persons with psychiatric disabilities must be considered.

Residential Status

Historically, the residential status of persons with psychiatric disabilities has been assessed in two ways: recidivism rates and place of residence. Hospital recidivism figures continue to be used routinely even though their use has been repeatedly critiqued on the grounds of their susceptibility to a variety of influences other than the persons' adjustment to their living situation (Anthony et al., 1972,1978; Bachrach, 1976) Comprehensive reviews of the literature indicate a gradually increasing rate of recidivism as the follow-up period lengthens. As mentioned in chapter 2, the

42

recidivism rate at 6 months is approximately 30–40%; at 12 months, 35–50%; and at 5 years, 65–75% (Anthony et al., 1972, 1978).

With respect to the place of residence, unfortunately most of the residential data cannot be compared across studies. Each study defines types of residences (e.g., group homes, transitional homes, foster homes) differently. Thus, there are no accurate estimates of the *typical* degree of independent living status of persons with severe psychiatric disabilities. A multistate sample of clients attending community support programs indicates that 40% live in a private home or apartment, 12% in board and care settings, 10% in family foster care placements, and the rest in other residential categories with less than 10% each (Tessler & Goldman, 1982). A later survey of a similar sample reported that 57% were living in private homes and apartments, with a significantly lower percentage in structured community settings (Mulkern & Manderscheid, 1989). The National Plan for the Chronically Mentally Ill (U.S. Department of Health and Human Services, 1980) estimated that of those persons residing in the community, 38–50% lived in board and care settings and 19–21% lived with families. A nationwide survey of family members of persons with severe psychiatric disabilities, who were also members of the National Alliance for the Mentally Ill (NAMI), reported that approximately 30% lived at home with family, 15% in a community residence, and 18% in hospitals (Spaniol et. al., 1984). It is clear from these studies that the type and definition of residential categories varies from study to study, thus precluding, for the time being, national estimates of independent living status.

Educational Status

The educational status of persons with severe psychiatric disabilities has received scant attention in the literature. However, the available data do point to the advanced educational status of many persons with severe psychiatric disabilities. Depending on the particular sample taken, between 52–92% of persons with severe psychiatric disabilities are high school graduates, and 15–60% of these high school graduates have attended college. For example, the NAMI survey data (Spaniol et al., 1984) indicated that of the total sample, 92% graduated from high school, 59% also attended college, and 17% graduated from college. Admittedly, the NAMI survey sampled middle to upper-middle income families, thus accounting for the relatively high educational levels, but data from the larger study of clients attending the Transitional Employment Program (TEP) at Fountain House in New York City indicated that approximately 70% were high school graduates, 48% also attended college, and 14% were college graduates (Fountain House, 1985). These figures are the same as the figures for the general population of New York City. A study of 505 persons with long-term psychiatric disabilities discharged from hospitals in Toronto, Canada, reported that 72% were high school graduates and 16% also attended college (Goering, Wasylenki, Lancee, & Freeman, 1984). Two studies of Community Support Program (CSP) clients reported 53–55% high school graduates and 19–23% with post-high school education (Mulkern & Manderscheid, 1989; Tessler & Goldman, 1982).

Vocational Status

Studies containing data on the competitive employment rate of persons discharged from psychiatric hospitals have been periodically surveyed by Anthony and associates (Anthony et al., 1972, 1978; Anthony, Howell, & Danley, 1984). As mentioned in chapter 2, the data have been fairly consistent, suggesting a full-time competitive employment figure of 20–30% for all persons discharged from psychiatric hospitals. However, if just persons with long-term psychiatric disabilities are studied, the full- and part-time competitive employment figure drops to approximately 15% and below. For example, the NAMI survey of well-educated persons with long-term psychiatric disabilities from middle- to upper-income families reported a full-time employment rate of about 5% (Spaniol et al., 1984). Farkas, Rogers, and Thurer (1987) followed up 54 long-term state hospital inpatients who had been targeted for deinstitutionalization in 1979. Over a 5-year period, 0% obtained competitive employment. Tessler and Goldman (1982) reported 11% full- and part-time competitive employment for CSP clients. The later survey by Mulkern and Manderscheid (1989) of CSP clients reported a competitive employment rate of slightly less than 10%. Wasylenki et al. (1985) reported that 11% of their hospitalized sample were employed prior to admission. Dion, Cohen, Anthony, and Waternaux (1988) followed patients hospitalized with bipolar disorder and found evidence of the impact of severity on vocational outcome. At 6-month follow-up, 64% of first admission patients were employed at some level of competitive employment versus 33% of persons with one or more prior hospitalizations. Interestingly, only 20% of the total sample were functioning at their expected level of employment, that is, at the occupational level expected on the basis of previous work history and education.

Review of the Literature

Empirical studies that met the inclusion criteria have been organized around the key dimensions of location, outcome measures, types of intervention, research design, and outcomes. Every effort was made to include all the relevant psychiatric rehabilitation research studies that appear in all the various mental health and rehabilitation journals. Of course, some relevant studies were unavailable at the time of this review because they were either in press, being prepared for journal submission, or had just been received by the granting agency that supported the research. Computer searches were conducted of several databases including Psychological Abstracts, Dissertation Abstracts, Mental Health Abstracts, REHABDATA, and ERIC. Current, not-yet-indexed journal issues were read, and major funding sources of psychiatric rehabilitation research (NIMH, National Institute on Disability and Rehabilitation Research) were contacted for information about recently completed studies. Most of the studies reviewed occurred within the last 10 years because few databased evaluations of psychiatric rehabilitation existed before the mid-1970s.

Location

Psychiatric rehabilitation studies have been conducted in a variety of settings, the most common being psychosocial rehabilitation centers and hospitals. Other settings have been community mental health centers and offices of state divisions of vocational rehabilitation. A number of studies have investigated the delivery of psychiatric rehabilitation interventions in more than one setting: hospital/community linkages (Paul & Lentz, 1977; Wasylenki et al., 1985) and mental health—vocational rehabilitation collaborative programming (Dellario, 1985; Rogers, Anthony, & Danley, 1989). Many of the single settings (e.g., psychosocial rehabilitation centers) use places outside the setting's physical location, such as employment sites (Fountain House, 1985). That the rehabilitation intervention often occurs in more than one place illustrates the need for a comprehensive and coordinated approach to planning service delivery.

As would be expected, much of the psychiatric rehabilitation research literature focuses on the vocational environment. In part, this is because the term *rehabilitation* emerged in a vocational context, that is, the vocational rehabilitation of persons with physical disabilities. However, within the mental health field, rehabilitation focuses on the residential environment as well; residential success and satisfaction is a factor closely related to the amelioration and aggravation of the psychiatric condition. This fact is reflected in the early and routine collection of recidivism data (Anthony et al., 1972; 1978). The research investigations reviewed are characteristic of the field's interest in both vocational and residential environments.

In contrast, as evidenced by the types of outcome measures used (e.g., recidivism), investigations of the improvement of educational status have been rare. A recent research project, investigating a university-based rehabilitation program for young adults, indicates the need for such educational interventions (Center for Psychiatric Rehabilitation, 1989; Unger, Danley, Kohn, & Hutchinson, 1987). The announcement of this program to a selected group of mental health agencies in the Boston area generated 75 completed applications for the first 26 available slots.

Outcome Measures

The field has moved beyond a simple reliance on recidivism and employment figures as outcome measures. In the first comprehensive review of the psychiatric rehabilitation field, Anthony, et al. (1972) had to rely exclusively on studies that reported recidivism and employment outcomes, the only types of data routinely collected. Yet, even these early studies suggested a lack of relationships between outcome measures across environments (in this instance, recidivism and employment) and the need for more refined measures in all the different outcomes of interest (Anthony et al., 1972).

In a comprehensive review of outcome measures used in psychiatric rehabilitation, Anthony and Farkas (1982) concluded that:

1. Change on a single measure of client outcome does not indicate that seemingly related measures of change have been affected (e.g., change in vocational functioning may not be correlated with change in psychosocial functioning).
2. A positive effect on one client outcome measure may have an associated negative effect on another (e.g., an increase in skills associated with an increase in anxiety).

Thus, it behooves researchers to study the impact of their intervention on a wider range of outcomes and to make no assumptions about the intervention's impact on outcomes not specifically studied. In general, the recent research literature seems compatible with these guidelines.

The types of outcomes used in the reviewed studies indicate the improvement in outcome measures. In terms of vocational status, simple yes/no measures of employment have been complemented by measures of types of employment, for example, prevocational, transitional, supported, part-time, full-time (Fountain House, 1985), earnings (e.g., Bond, 1984), satisfaction with work (National Institute of Handicapped Research, 1980), productivity (Hoffman, 1980), and instrumental role functioning (Goering, Farkas, Wasylenki, Lancee, & Ballantyne, 1988). With respect to residential status, simple recidivism measures have been all but replaced by measures of total days in the community (e.g., Cannady, 1982), social adjustment (e.g., Linn, Caffey, Klett, Hogarty, & Lamb, 1979), number of friends and activities (e.g., National Institute of Handicapped Research, 1980; Vitalo, 1979), degree of independent living (e.g., Mosher & Menn, 1978), satisfaction with community adjustment (e.g., Katz-Garris, McCue, Garris, & Herring, 1983), and social skills (e.g., Aveni & Upper, 1976).

Few studies investigate educational status, yet many academically capable young adults have had their educational progress interrupted by a psychiatric disability (Spaniol et al., 1984; Unger & Anthony, 1984). Unger and Anthony (1989) have reported on a year-long, university-based, supported education program for young adults who are severely psychiatrically disabled. Results indicated that approximately 75% completed the program, and that throughout the follow-up period of 3 years, approximately 45-50% were working or in school. The outcome of rehabilitation interventions designed to improve functioning in the educational area could be assessed by using such outcome measures as degree programs entered, degree programs completed, courses completed, professional/educational certificate programs completed, academic skills learned, course grades, and achievement test scores.

Type of Intervention

In order to be included in the review, the intervention had to provide either client skill development, increased support, or both. However, several obvious limitations must be noted in how psychiatric rehabilitation interventions have been studied. First, many of the interventions are not described in sufficient detail to permit replication, either in future research studies or in clinical practice. Simply knowing the setting of the intervention provides little information about the specific intervention. Limitations imposed by journal space no doubt account for some of this brevity, but even the references accompanying most articles do not indicate the existence of any materials (e.g., manuals or videotapes) that might facilitate replication. Exceptions are the interventions modeled after the psychiatric rehabilitation interventions developed by Anthony and his colleagues (see for example, Cohen, Danley, & Nemec, 1985; Cohen, Farkas, & Cohen, 1986; Cohen, Nemec, Farkas, & Forbess, 1989; Cohen, Farkas, Cohen, & Unger, 1990; Goering, Wasylenki et al., 1988; National Institute of Handicapped Research, 1980; Vitalo, 1979; Wasylenki et al., 1985); the Fountain House transitional employment approach

(Fountain House, 1985); Liberman's social skills training (Liberman, Mueser, & Wallace, 1986); Paul's social learning approach (Paul & Lentz, 1977); and Azrin's job seeking skills program (Jacobs et al., 1984).

A second problem is the difficulty in separating the unique contributions to rehabilitation outcome of skill development versus support interventions. Support interventions typically mean supportive programs that accommodate the persons' deficits (e.g., sheltered workshops) or supportive persons that help the person to meet the demands of nonsupportive environments (e.g., these supportive persons provide personal counseling, companionship, advocacy, and practical advice). Some of these supportive programs and people also provide opportunities for persons with severe psychiatric disabilities to learn skills. Some programs may promote skill development by providing an environment that facilitates learning; other programs provide a more structured, formalized skill teaching program. From a research perspective, skill development and support interventions have been inextricably linked. Investigations have lacked either the technology or the intent to study the relative merits of the two types of interventions. The research currently suggests that interventions that provide the client with an opportunity both to learn skills and to receive support remain the preferred rehabilitation interventions.

Research Design

Just as the types of outcomes studied have become more refined, the research designs have become more rigorous—albeit more slowly. Studies already exist, and are being done more frequently, that use random assignments to experimental and control groups (Bond, 1984; Dincin & Witheridge, 1982; Paul & Lentz, 1977; Wolkon, Karmen, & Tanaka, 1971; Ryan & Bell, 1985; Bond & Dincin, 1986) or matched experimental and control groups (Wasylenki et al., 1985; Goering, Wasylenki, et al., 1988; Vitalo, 1979; Beard, Malamud, & Rossman, 1978; Hoffman, 1980; Mosher & Menn, 1978; Matthews, 1979). The positive outcomes of these studies seem in no way different from the outcomes generated by quasi-experimental designs.

However, the design deficiencies in the psychiatric rehabilitation research literature are readily apparent. Few studies have designs that permit reasonable, causal inferences to be made. Also, many designs are plagued by inadequate sample size, heterogeneity of population, nonrandom assignment, lack of specificity and replicability of treatment approach, and lack of outcome measures appropriate to the interventions used. In addition, the vast majority of studies use one group and posttest only designs.

Yet these nonexperimental studies still do have value. They have provided the empirical and conceptual foundation for the experimental studies that have begun to appear. They have seized the available data and used the results to fashion more specific interventions that can now be researched experimentally. Much of this previous research can be considered exploratory, examining the practical significance of interventions before the experimental test. When viewed from this perspective, the psychiatric rehabilitation field is ready for the additional, critical, experimental studies that need to be undertaken.

Outcomes

Any discussion of outcomes is, of course, limited by the previously discussed design deficiencies. However, as a group the research studies suggest that rehabilitation interventions produce rehabilitation outcomes in the clients served. These studies can be grouped by the specific types of rehabilitation programs that were evaluated.

Several studies reported positive outcomes through joint programming between two different settings or agencies. These coordinated programs included cooperative hospital and community programming (Becker & Bayer, 1975; Jacobs & Trick, 1974; Paul & Lentz, 1977; Wasylenki et al., 1985) and collaborative mental health and vocational rehabilitation programs (Dellario, 1985; Rogers, Anthony, & Danley, 1989). That collaborative interventions initiated from a hospital base can produce rehabilitation outcome in the community is particularly noteworthy. As mentioned in chapter 2, Dellario and Anthony (1981), based on their own literature review as well as on reviews by Kiesler (1982) and Test and Stein (1978), concluded that hospital care and community-based care should not be compared with one another but rather to the stated mission of the agency, no matter what its location. This, of course, presumes a statement of mission that articulates the intended outcomes for each facility—a condition that apparently does not exist in many settings (Farkas, Cohen, & Nemec, 1988). Nevertheless, replacing hospital care with community care requires a commitment to provide rehabilitation programming without arbitrary time limits. The critical determinant of the effectiveness of services provided within a particular setting may well be the relationship of that setting to an overall system of long-term community support. Even though use of public psychiatric hospitals continues to decline (Bachrach, 1986a, 1986b), and the vast majority of public mental health systems continue to emphasize community-based services, psychiatric hospitalization per se remains a significant part of public mental health systems. The most dramatic changes have been in the increased use of the general hospital setting and the growth of private psychiatric hospitals. These trends present potential problems in providing relevant services, given the often limited experience of these new settings with long-term psychiatric disabilities, the psychiatric rehabilitation approach, and community support systems.

While policy debates over the role of the hospital in mental health systems continue, change to a rehabilitation orientation in at least some hospital settings is promising. The technology of changing a traditional treatment program to a psychiatric rehabilitation program is available (Anthony, Cohen, & Farkas, 1987; Farkas, Cohen, & Nemec, 1988). The research studies suggest that a rehabilitation approach that provides integrated development of skills and supports within the variety of settings in which clients are served may be the most effective approach.

The positive vocational outcomes associated with collaborative mental health and vocational rehabilitation interventions suggest the benefits of better coordination among existing services. In an era of cost containment, such data give impetus to increasing the effectiveness of those service components already in place. More efficient service delivery need not sacrifice improved client outcome—indeed, it may improve it.

With the continued development of psychosocial rehabilitation centers, research conducted at these centers takes on increasing importance. Several studies conducted at psychosocial rehabilita-

51

tion centers are studies of a Transitional Employment Program (TEP). TEP is a vocational training innovation currently in use in many psychosocial rehabilitation agencies around the country (Fountain House, 1985). In a traditional TEP, a member (client) of the psychosocial rehabilitation center is placed in an entry-level job in a normal place of business. All placements are temporary (3–9 months), typically half-time, and are supervised by the psychosocial rehabilitation center (Beard et al., 1982). A TEP is designed to develop the self-confidence, job references, and work habits necessary to secure permanent employment.

Until recently, there has been little research studying the effectiveness of TEPs. Typically, studies have examined the vocational outcomes of persons served by a psychosocial rehabilitation center, whose services included a TEP. In a very early study, Beard, Pitt, Fisher, & Goertzel (1963) reported no significant differences in employment between experimental and comparison subjects. In a randomized control group design, Dincin and Witheridge (1982) found no differences in employment at the 9-month follow-up.

More recently, a major follow-up study of Fountain House members who spent at least one day in a TEP has been completed (Fountain House, 1985). The results for 527 individuals who participated in a TEP indicated that employment outcome increased as a function of the time since the initial TEP. For members whose initial TEP was at least 42 months ago, 36% were competitively employed. At 12 months and 24 months, the employment rates were 11% and 19%, respectively.

Thresholds, a psychosocial rehabilitation center, conducted another study of two different types of TEPs, an accelerated TEP and a traditional TEP. Persons in the traditional TEP were required to remain in prevocational settings a minimum of 4 months longer than persons in the accelerated TEP condition. At the 15-month follow-up, the 20% and 7% employment rates in the accelerated and traditional TEP conditions approximate Fountain House's employment figure for the same time period (Bond & Dincin, 1986).

The results of TEP research conducted by Fountain House suggest that TEPs have a significant impact on employment as the follow-up period increases. The recent research conducted by Thresholds suggests that for a person with prior work experience, the entrance into a TEP could be accelerated and the time needed to obtain employment shortened.

Psychiatric rehabilitation outcome studies are somewhat atypical in the mental health field in that many of the intended outcomes are specific, observable, understandable, and valued by the general public. Various methods of skill development and/ or support interventions have been found to have an impact on such seemingly straightforward outcome criteria as days in the community (Cannady, 1982; Paul & Lentz, 1977), earnings (Bond, 1984), reduction in disability pensions (Jensen, Spangaard, Juel-Neilsen, & Voag, 1978), and employment (Turkat & Buzzell, 1983). Other outcomes produced by rehabilitation, although somewhat more complex in terms of measurement, have meaning to the lay person, such as independent living (Bond, 1984; Mosher & Menn, 1978), productivity (Ryan & Bell, 1985), work satisfaction (National Institute of Handicapped Research, 1980), increased friendships and activities (Vitalo, 1979), role performance (Goering, Farkas, et al., 1988; Jacobs & Trick, 1974), and increased skills (LaPaglia, 1981; Vitalo, 1979). As the reviewed research studies suggest, psychiatric rehabilitation interventions are typically very specific in terms of their intended outcomes.

As outcome measurements in psychiatric rehabilitation studies have become more refined, i.e., measuring degrees of improvement over time rather than yes/no categorical measures at one point in time, the findings with respect to psychiatric rehabilitation interventions have become more positive. For example, studies that have not shown the expected change on recidivism did report results on measures of community tenure (Beard et al., 1978) and instrumental role functioning (Goering, et al., 1988).

In several studies, the longer the follow-up period, the more dramatic the findings. For example, differences in instrumental role functioning and social adjustment increased from the 6-month to the 2-year follow-up (Goering, et al., 1988); differences in employment not apparent at 9 months appeared at 15 months (Bond & Dincin, 1986); and the longer the transitional employment placement, the greater the vocational outcome (Fountain House, 1985).

Conclusions

Within the obvious limitations of the measurement, design, and description of the intervention, current research suggests that a psychiatric rehabilitation approach does affect rehabilitation outcome positively. Almost all of the studies combine different elements of skill development and support, making it impossible at this time to unravel the unique contributions of the different elements to outcome. In 1974, Anthony and Margules conducted a review of the literature and tentatively concluded that persons with severe psychiatric disabilities can learn important skills despite their symptomatology, and that these skills, when combined with appropriate community supports, can have an impact on rehabilitation outcome. A decade later the same conclusion can be drawn, based on additional data and relatively more sophisticated research designs.

Future Research Issues

Monumental problems still exist in psychiatric rehabilitation research. The most critical problem is the continued need for experimental research on replicable, rehabilitation interventions. Mosher and Keith (1979), Goldberg (1980), Meyerson and Herman (1983) and Keith and Matthews (1984) have all called for well-controlled process and outcome studies of a psychiatric rehabilitation approach.

Further research of psychiatric rehabilitation is critical because community mental health centers are now expanding their services in a renewed attempt to treat persons with long-term psychiatric disabilities in the community. A NIMH-sponsored study by Larsen (1987) has indicated that the mental health services currently experiencing greatest growth are those primarily directed toward persons with long-term psychiatric disabilities in the community—that is, those persons most apt to receive a psychiatric rehabilitation approach. The high percentage of treatment dollars used to serve persons with long-term psychiatric disabilities has been well documented, and more and more mental health authorities are developing psychiatric rehabilitation services for this population. The psychiatric rehabilitation approach promises to gain widespread acceptance and to have pragmatic relevance to the needs of persons with severe psychiatric disabilities (e.g., surveys by Lecklitner & Greenberg, 1983; Spaniol, Jung, Zipple, & Fitzgerald, 1987). Trends such as this make standardized, well-controlled psychiatric rehabilitation research even more important.

One of the most difficult problems to overcome in future psychiatric rehabilitation outcome research is the development of practical, meaningful, reliable, and valid outcome measures. Many measures, particularly those adopted from the social skills training literature, reflect molecular behavior. Moving from such measures as frequency counts of the number of eye contacts made in 3 minutes of conversation to the more complex assessment of social behavior involves a loss of clarity and specificity. Two common responses to the problem have been either to translate more complex social behavior into molecular behaviors (Wallace et al., 1980) or to develop idiosyncratic measures for each new set of research problems. Because of the limited space in professional journals, instruments are not often described in sufficient detail. Consequently, replicating the measurement can be as difficult as replicating the rehabilitation interventions, which are also incompletely described. Many studies do not even reference the instruments that were used, making it more difficult to identify the potential standardized measures.

Another problem that must be corrected in future research studies is that reliability and validity measures of instruments are often not reported (Ciminero, Calhoun, & Adams, 1977). Reliability tends to be more frequently reported than validity—perhaps reflecting the problems in establishing convergent or concurrent validity for a vaguely defined set of complex behaviors (Alevizos & Callahan, 1977). Indeed, the question of the validity of behavioral measures has rarely been considered (Wallace et al., 1980). For community adjustment measures, however, even formal tests of reliability are seldom documented (Waskow & Parloff, 1975). Many instruments used in outcome research were developed for the assessment of the acute population and are applied to the long-term population, for lack of a population-specific instrument. When the researcher moves beyond direct measurements of change (e.g., number of days attending a workshop) and applies assessment techniques standardized on other populations, problems with reliability and validity arise.

Researchers are unanimous in calling for future research measures designed to assess the generalization of skill acquisition. Hersen and Bellack (1977, p. 510) stated that the "importance of generalization in most behavioral research has been acknowledged more in print than in actual practice." There is an acknowledged lack of a systematic framework in which to produce generalization techniques or measures that are ethical and naturalistic (Curran, 1980; Hersen & Bellack, 1977). Only 40% of studies targeting the skills of persons with long-term disabilities addressed some aspect of the problem of skill application (Wallace et al., 1980). In those studies that have addressed generalization, frequency counts of behaviors and unstructured or structured questionnaires have typically been the assessment measure used (Hollingsworth & Foreyt, 1975; Patterson & Teigen, 1973; Tracey, Briddell, & Wilson, 1974).

In addition to assessing whether the skills learned in training are applied in the targeted environment, there is a pressing need for the development of skill measures that positively correlate with concrete measures of client benefits. Skill outcome measures developed by Paul and Lentz (1977) and by Griffiths (1973) are models in this regard. Predictive validity based on measures of inpatient behaviors has been reported by Power (1979) and Redfield (1979) using the Clinical Frequencies Recording System and the Time Sample Behavior Checklist, respectively. Both instruments were developed and standardized in the decade-long investigation conducted by Paul and his associates. Predischarge scores on these instruments can significantly predict the discharged patient's functioning in the community (Paul, 1984).

Designed to measure vocational behaviors of persons with psychiatric disabilities, the Standardized Assessment of Work Behavior (Griffiths, 1973, 1974; Watts, 1978) assesses a broad range of behaviors (e.g., uses tools/equipment, communicates spontaneously, grasps instructions quickly). Items are rated on a continuum from a strength (e.g., looks for more work) to a deficit (e.g., waits to be given work). Both reliability and predictive validity data are available for this scale.

Recently, interest has increased in outcome measures of

overall level of functioning, compatible with the increased use of functional assessment in psychiatric rehabilitation practice (Cohen & Anthony, 1984). Many states have developed instruments for measuring the functional levels of their Community Support Program clients. These instruments often include ratings of client skills. The CSS-100 (New York State Office of Mental Health, 1979) is used by many community support systems. Separate scales measure adjustment to an environment (e.g., using public transportation, managing funds, dressing self) and behavioral problems or symptoms (e.g., hospitalization, employment-related services, community living programs, socialization activities). Similar to the New York State instruments, the Multi-Function Needs Assessment (Angelini, Potthof, & Goldblatt, 1980) used in Rhode Island and Connecticut includes an assessment of functioning, self-maintenance, environmental interactions, psychiatric symptoms, and current use of services. Other instruments, developed along the same lines, are in use in New Jersey (New Jersey Division of Mental Health and Hospitals, 1980) and Michigan (Cornhill Associates, 1980). These instruments have shown promise for outcome studies with a large number of subjects (e.g., all patients in the hospital, all mentally ill clients served by the state). Although these instruments represent a first step in grappling with the problems of measuring changes which result from psychiatric rehabilitation interventions, they are, nevertheless, still crude. Ridgway (1988), in reviewing over 200 level of functioning instruments with respect to residential needs, points out the lack of specificity in the items and their lack of relevancy to client's preferred housing settings as major deficits in the validity and meaningfulness of these standardized instruments.

Summary

Large-scale experimental studies of psychiatric rehabilitation are not only needed, they are also increasingly feasible. Issues of measurement are being addressed. Currently, a number of psychiatric rehabilitation interventions can be described at a level of detail that permits their implementation to be observed and monitored reliably. Thus, the researchers can collect data as to the degree to which the intervention under investigation was implemented. The intervention is now sufficiently described so that, if the results are promising, it can be replicated in service settings and clinical research programs.

In summary, the reviewed research on innovative rehabilitation programs has illustrated that the stage is set for experimental studies of replicable, measurable psychiatric rehabilitation interventions.

Philosophy

4

The essence of a metaphor is understanding and experiencing one kind of thing in terms of another.... We see metaphor as essential to human understanding and as a mechanism for creating new meaning and new realities in our lives.

Lakoff & Johnson

The underlying philosophy of psychiatric rehabilitation can be made more understandable by using a metaphor or analogy. The practice of rehabilitation with persons who are physically disabled can serve as an analogy for the practice of rehabilitation with persons who are psychiatrically disabled. Explaining psychiatric rehabilitation by using an analogy to physical rehabilitation (e.g., rehabilitating persons with spinal cord injury or cardiac disease) makes psychiatric rehabilitation more comprehensible. An analogy is much more easily remembered than is a definition or a list of concepts. The image creates understanding. The image of physical rehabilitation can be used to create

Parts of this chapter were excerpted with permission from:

Anthony, W. A., Cohen, M. R., & Cohen, B. F. (1984). Psychiatric rehabilitation. In J. A. Talbott (Ed.), *The chronic mental patient: Five years later* (pp. 137–157). Orlando, FL: Grune & Stratton.

Anthony, W. A., & Liberman, R. P. (1986). The practice of psychiatric rehabilitation: Historical, conceptual, and research base. *Schizophrenia Bulletin*, *12*, 542–559.

understanding of psychiatric rehabilitation and to inspire commitment to psychiatric rehabilitation. In our society rehabilitation of persons who are physically disabled is very much valued. Can society do less for persons who are psychiatrically disabled? This analogy must be clear in the public's mind and be used to further the public's understanding of psychiatric rehabilitation.

The analogy to physical rehabilitation can also clarify the meaning of psychiatric rehabilitation for the many different disciplines that are involved in its practice. Even though obvious differences exist between a psychiatric disability and a physical disability, enough similarities between the two fields remain. For example, persons with physical disabilities and persons with psychiatric disabilities both exhibit handicaps in role performances; need a wide range of services, frequently for an extended period of time; and may or may not experience a total recovery from their disabilities.

There are several other advantages in using physical rehabilitation as an analogy for explaining psychiatric rehabilitation. A physical disability is considered to be less stigmatizing than is a psychiatric disability. Furthermore, the rehabilitation of persons with physical disabilities appears more credible and understandable to the layperson. Thus, by using the analogy in talking about psychiatric disability, the field of psychiatric rehabilitation becomes more legitimate and acceptable.

Presently, mental health professionals are beginning to understand how the philosophy of rehabilitation may be relevant to persons with severe psychiatric disabilities. However, at present the trained personnel and effective programs needed to put this philosophy into practice are not typically available in communities across the country.

Rehabilitation Model

Within the last decade, an increasing consensus has developed about the underlying philosophy of rehabilitation. Table 4–1 illustrates the rehabilitation model on which psychiatric rehabilitation philosophy is based.

The basic concepts of impairment, disability, and handicap

in the rehabilitation model have been described somewhat differently over the years. However, the integrative work of Wood (1980) and Frey (1984) has brought conceptual clarity to these terms. As depicted in Table 4–1, the impairment of structure or function can lead to a decreased ability to perform certain skills and activities, which in turn can limit the person's fulfillment of certain roles.

Historically, mental health treatment has intervened at the impairment stage. Somatic and psychological treatment attempts to alleviate the signs and symptoms of pathology. As mentioned in chapter 2, Leitner and Drasgow (1972), in analyzing the difference between treatment and rehabilitation, point out that, in general, treatment is directed more toward minimizing sickness and rehabilitation more toward maximizing health. Eliminating or suppressing an impairment (i.e., treatment) does *not* lead automatically to more functional behavior (i.e., reducing the disability). Likewise, a decrease in disability does not automatically lead to reductions in impairment, although such a possibility could occur (Strauss, 1986). Significantly, chronic or severe impairment (e.g., diabetes, stroke) does *not* always mean a chronic disability and handicap. What the impairment does is to increase the risk of chronic disability or handicap.

The clinical practice of psychiatric rehabilitation, just like its counterpart in physical rehabilitation, is comprised of two intervention strategies: (1) client skill development, and (2) environmental support development. Psychiatric rehabilitation practice is guided by the basic philosophy of rehabilitation: disabled persons need *skills* and *environmental supports* to fulfill the role demands of their living, learning, social, and working environments. The assumption of clinical rehabilitation is that by changing their use of skills and/or the supports in their immediate environments, persons with psychiatric disabilities will be more able to perform those activities necessary to function in specific roles of their choice. In other words, interventions designed to lessen the disability are assumed to lead to a decrease in handicap.

In addition to clinical rehabilitation interventions, persons with psychiatric disabilities can be helped to overcome their handicaps through societal rehabilitation interventions (Anthony, 1972). Societal rehabilitation is designed to change the system in which

TABLE 4–1 *The Rehabilitation Model*

Stages:	I. Impairment	II. Disability	III. Handicap
Definitions:	Any loss or abnormality of psychological, physiological, or anatomical structure or function.	Any restriction or lack of ability to perform an activity in the manner or within the range considered normal for a human being. (resulting from an impairment)	A disadvantage for a given individual that limits or prevents the fulfillment of a role that is normal (depending on age, sex, social, cultural factors) for that individual. (resulting from an impairment and/or a disability)
Examples:	Hallucinations, delusions, depression,	Lack of: work adjustment skills, social skills, ADL skills	Unemployment, homelessness
Typical Services:	Treatment focused on alleviating or eliminating pathology.	Clinical rehabilitation focused on developing client skills and environmental supports.	Societal rehabilitation focused on changing the system in which the individual lives.

Adapted from: Center for Psychiatric Rehabilitation (1984) *Annual report for the National Institue of Handicapped Research.* Boston: Boston University.

psychiatrically disabled persons must function. Unlike clinical rehabilitation, its focus is not on the unique skills and supports of specific individuals. Rather, the focus is on changing society to help the class of persons with psychiatric disabilities become more successful and satisfied in environments of their choice. Examples of societal rehabilitation interventions are the Targeted Job Tax Credit legislation, changes in the work incentives within the Social Security Disability program, and the development of a European-type quota system for the employment of disabled workers. The importance of these societal interventions is a strong reminder that overcoming a handicap may be more a function of a non-accommodating and discriminating social and economic system than it is a function of the person's impairment and disability (Rosen, 1985).

Clinical and societal rehabilitation interventions are not mutually exclusive (Stubbins, 1982). In fact, the 1973 amendments to the Vocational Rehabilitation Act (which mandates clinical rehabilitation interventions) also recognized the value of societal rehabilitation. The amendments established the principle of affirmative action by contractors who do business with the federal government and also attempted to establish the government as a model employer with respect to architectural access (Stubbins, 1982). Furthermore, just as clinical and societal rehabilitation interventions are complementary, a rehabilitation intervention can be seen as complementary to treatment interventions. A person who suffers severe psychiatric problems is usually in need of both treatment and rehabilitation interventions.

Basic Principles of Psychiatric Rehabilitation

As the rehabilitation model has gained greater acceptance, the principles underlying this conceptual model have become clearer (Wright, 1981). These principles operate in a number of model programs designed to rehabilitate persons with severe psychiatric disabilities (Bachrach, 1989). Descriptions of model psychiatric rehabilitation programs are appearing with increasing regularity in the literature (e.g., see Bachrach, 1980; 1988b). Because of the current wealth of articles about psychiatric rehabilitation, it is now possible to articulate a set of principles basic to most rehabilitation settings. The specification of basic psychiatric rehabilitation principles has heretofore been a difficult task because psychiatric rehabilitation is practiced in many diverse service settings by practitioners from different professional disciplines. The identification of a set of principles basic to psychiatric rehabilitation can highlight the commonalities among diverse psychiatric rehabilitation settings and the different disciplines practiced within them (Cnaan, Blankertz, Messinger, & Gardner, 1988). The writings of Beard (Beard et al., 1982), Lamb (1982), Dincin (1981), Grob (1983), and Anthony (1982), among others, indicate a growing consensus about what constitutes the essential principles of psychiatric rehabilitation.

The following principles seem to be basic to the practice of psychiatric rehabilitation and operate independently of both the settings in which they are practiced and the professional disciplines of practitioners. Table 4–2 lists nine basic principles of psychiatric rehabilitation.

Nine Principles

1. The primary focus of psychiatric rehabilitation is on improving the competencies of persons with psychiatric disabilities. As suggested previously, rehabilitation is directed primarily at maximizing health rather than minimizing sickness, that is, health induction, not simply symptom reduction. The historical emphasis in physical rehabilitation has always been on increasing competencies rather than on simply alleviating symptoms and pathology. It is a well-known rehabilitation axiom that minimizing or suppressing sickness will not automatically lead to improvement in functional capacity. Put another way, the emphasis is on *coping* rather than *succumbing*, that is, on the challenge for meaningful adaptation or change rather than on the difficulties and heartbreak of the disability (Wright, 1980). Or, as Beard has indicated, a fundamental message of rehabilitation is the "belief in the potential productivity of the most severely psychiatrically disabled client" (Beard, et al., 1982, p. 47). Dincin (1981) speaks of psychiatric rehabilitation as providing growth-inducing opportunities. Al-

TABLE 4–2   *Basic Principles of Psychiatric Rehabilitation*

1. The primary focus of psychiatric rehabilitation is on improving the competencies of persons with psychiatric disabilities.

2. The benefits of psychiatric rehabilitation for the clients are behavioral improvements in their environments of need.

3. Psychiatric rehabilitation is eclectic in the use of a variety of techniques.

4. A central focus of psychiatric rehabilitation is on improving vocational outcome for persons with psychiatric disabilities.

5. Hope is an essential ingredient of the rehabilitation process.

6. The deliberate increase in client dependency can lead to an eventual increase in the client's independent functioning.

7. Active involvement of clients in their rehabilitation process is desirable.

8. The two fundamental interventions of psychiatric rehabilitation are the development of client skills and the development of environmental supports.

9. Long-term drug treatment is an often necessary but rarely sufficient component of a rehabilitation intervention.

though the specific words used to convey the message of this principle are numerous (i.e., growth, productivity, health, coping, skills, competencies, capabilities), the meaning of the principle remains clear.

2. The benefits of psychiatric rehabilitation for the clients are behavioral improvements in their environments of need. The development of therapeutic insight is not a primary goal (Anthony, 1982; Dincin, 1981). The emphasis is on reality factors rather than intrapsychic factors (Lamb, 1982), on improving the person's ability to do something in a specific environment. Despite the presence of residual disability, rehabilitation attempts to help the person adjust and adapt to the requirements of specific environments (Grob, 1983). Psychiatric rehabilitation focuses on the person's ability to perform certain behaviors within certain environments, a focus similar to the emphasis in physical rehabilitation. For example, outcome for a person who is blind is not just learning to move around without the aid of sight; it is also the application of these mobility skills in certain environments of need (e.g., home, work). hus, client outcome is tied to an environment. Similarly, the psychiatric rehabilitation specialist must not just work toward improving skills (e.g., conversational skills), but must do so with respect to the demands of the specific environment in which the client is presently functioning or will be functioning in the future (e.g., community residence, transitional employment placement).

3. Psychiatric rehabilitation is eclectic in the use of a variety of techniques. The philosophy of rehabilitation guides the practice of psychiatric rehabilitation; no allegiance exists to any personality or psychotherapeutic theory. Grob (1983, p. 278) has characterized psychiatric rehabilitation as "eclectic in theory and pragmatic in adaptation." Dincin (1981) indicates that the practice of psychiatric rehabilitation is not dependent on the acceptance of a particular theory as to the cause of mental illness. The practice of psychiatric rehabilitation is free to incorporate any technique that is effective for the intended purpose.

4. A central focus of psychiatric rehabilitation is on improving vocational outcome for persons with psychiatric disabilities. As Brooks (1981) indicated, work is the very essence of rehabilitation. The centrality of vocational functioning in psychiatric rehabilitation is most strongly apparent in the Fountain House approach. Work is a central focus in the Fountain House approach and underlies all aspects of this approach. The developers of the Fountain House approach believe "that work, especially the opportunity to aspire to and achieve gainful employment, is a deeply generative and integrative force in the life of every human being" (Beard et. al., 1982, p. 47). Grob (1983) holds that the appropriate job placement of the employable ex-patient is an essential part of the recovery process. Lamb (1982, p. 7) states that "work therapy geared to the capabilities of the individual patient should be a cornerstone of community treatment of long-term patients." The field of psychiatric rehabilitation is clearly rooted in a belief in the critical importance of work or worklike activities in the rehabilitation of persons with psychiatric disabilities (Cnaan et al., 1988; Connors, Graham, & Pulso, 1987; Harding, Strauss, Hafez, & Lieberman, 1987). Not only is the value of work reflected in rehabilitation principles, but vocational programming is also an integral part of the development of the field of psychiatric rehabilitation.

5. Hope is an essential ingredient of psychiatric rehabilitation. Rehabilitation is future oriented, and the tasks of the present are guided by a renewed and revived sense of hope for the future. Dincin (1981) suggests that the atmosphere of a psychiatric rehabilitation setting must be pervaded by a sense of hopefulness and an orientation towards the future. Fountain House attempts to instill a "helpful view of the future" (Beard, et al., 1982, p. 49).

Practitioners of both physical rehabilitation (Wright, 1960) and psychotherapy (Frank, 1981) have recognized the importance of hope or positive expectations for improvement as a critical factor in the recovery process (Deegan, 1988). Anthony, Cohen, and Cohen (1983) maintain that the hope for client improvement, no matter what the statistical probabilities of improvement are, is an important component of any rehabilitation intervention. In contrast, the absence of hope for improvement makes the practice of psychiatric rehabilitation problematic. When hopelessness permeates a setting and contaminates the attitudes of practitioners, the interminable and difficult demands of psychiatric rehabilitation practice seem overwhelming to practitioners, and the potential for client gain is diminished.

As mentioned previously, chronic or severe impairment (e.g., diabetes, stroke, arthritis) does not automatically lead to chronic disability and handicap. It only increases the risks of chronic disability and handicap in certain areas of functioning. For example, a person with paraplegia is disabled in terms of walking and handicapped in terms of being a firefighter, but is not disabled in terms of thinking nor handicapped as a computer programmer. A person with a psychiatric disability is not disabled for all time in all areas of functioning, nor handicapped in all roles. The belief in rehabilitation is that recovery in some, many, or all areas of functioning is possible.

Hope is always necessary but often insufficient. The hopeful stance must be combined with an ever-expanding psychiatric rehabilitation technology. (As Benjamin Franklin said, "If you live by hope alone, you will probably die fasting!") Hope and advances in technology are intimately related. Hope begets new technology and new technology begets new hope.

6. The deliberate increase in client dependency can lead to an eventual increase in the client's independent functioning. Supported or sheltered residential, educational, social, and vocational settings that allow for greater levels of client dependency are the traditional settings in which rehabilitation interventions occur (Lamb, 1982). Although the danger of overdependency must be recognized, the psychiatric rehabilitation philosophy distinguishes between types of dependencies (Havens, 1967). Dependency on a limited number of mental health personnel and settings is a natural first step in rehabilitation and in itself is not inherently destructive (Dincin, 1981).

Occasionally, in some mental health programs independence becomes so valued that client dependence becomes devalued. Yet from a rehabilitation perspective, dependency is not a dirty word (Anthony, 1982). Interventions with persons who are physically disabled often encourage dependency on persons or things in one environment so the client can function more effectively in another environment (Kerr & Meyerson, 1987). Dependency in one area of functioning can set a client free in another area (Peters, 1985). For example, dependence on a personal care attendant for help in dressing for work may allow a person with quadriplegia to hold a full-time job. Dependency in physical rehabilitation is a matter of degree, varying naturally between and within environments.

Persons with physical impairments are allowed a healthy dependence on people and things in their environment. For example, persons who are quadriplegic are not criticized for being dependent on a wheelchair! Society has made accommodations for commonplace physical dependencies. Imagine the controversy if it was suddenly ruled that only people with uncorrected 20/20 vision could drive, that is, persons who are dependent on eyeglasses would be prohibited from obtaining a driver's license.

The technology of psychiatric rehabilitation is limited in its ability to achieve total client independence from professional caregivers. Furthermore, dependence on people, places, activities, or things is a normal state of affairs. Interventions that allow for a certain degree of dependency at certain times, for example, the use of enablers or aides, may in fact be maximizing the client's functioning in other environments at other times (Weinman & Kleiner, 1978).

7. Active involvement of clients in their rehabilitation is desirable. Client involvement in psychiatric rehabilitation is the active participation by clients (e.g., communication of their values, experiences, feelings, ideas, and goals) throughout all the phases of rehabilitation. The research literature suggests that the rehabilitation goals and assessments of clients and practitioners frequently differ (Dellario, Goldfield, Farkas, & Cohen, 1984; Makas, 1980). The literature also confirms that helping (Carkhuff, 1969) and teaching (Aspy & Roebuck, 1977) is difficult without client or student involvement. Psychiatric rehabilitation incorporates the client's perspective in the assessment and intervention phases as well as the practitioner's perspective (Anthony, Cohen, & Farkas, 1982).

Client participation can also include involvement of clients in the planning and delivery of services to other clients. (Consumer-run mental health services are highlighted in chapter 8.) Fountain House has conducted a member (client) training program designed to increase and enrich the various roles that members assume in the Fountain House approach, including outreach services to other members, member education and tutoring, advocacy, and evaluation studies (Beard et al., 1982).

Client involvement requires rehabilitation procedures that can be explained to and understood by the client. The rehabilitation intervention cannot seem mysterious to the client. The practitioner must constantly try to demystify rehabilitation. Most critical to client involvement are the practitioner's commitment to the goal of client involvement and the practitioner's belief that rehabilitation is done *with* clients and not *to* clients. People do not *get* rehabilitated. They must be "active and courageous participants in their own rehabilitation" (Deegan, 1988, p. 12).

8. The two fundamental interventions of psychiatric rehabilitation are the development of client skills and the development of environmental supports. Interventions that attempt to improve either the person or the person's environment are the time-tested double focus of rehabilitation—be it physical rehabilitation or psychiatric rehabilitation (Wright, 1980). The focus on changing the person typically involves clients learning the specific skills they need in order to function more effectively in their environments. The focus on environmental change typically involves modifying the environment to accommodate or support the client's present skill functioning.

Practitioners of physical rehabilitation have traditionally used both types of interventions with their clients. For example, the physical therapist teaches the person with paraplegia new skills (e.g., how to make a wheelchair-to-bed transfer) or changes the person's environment to better accommodate the person's present skill level (e.g., ramps, wheelchair-accessible bathrooms).

Similar to the approach used in physical rehabilitation, the rehabilitation approach with persons who are psychiatrically disabled also focuses on building skills and modifying environments. The psychiatric rehabilitation approach is based on the research literature that has indicated that a person's skills, not symptoms, relate most strongly to rehabilitation outcome (Anthony & Margules, 1974). In addition, rehabilitation research has shown repeatedly that persons with psychiatric disabilities can learn a variety of physical, emotional, and intellectual skills regardless of their symptomatology. Furthermore, these skills, when properly integrated with support for the use of these skills in the community, can have significant impact on the clients' rehabilitation outcome (Anthony, 1979; Dion & Anthony, 1987).

Psychiatric rehabilitation settings vary in how systematically they approach skill building or environmental modification. Skill building and environmental modification may be informal and experiential (Beard et al., 1982; Dincin, 1981), or planned and systematic (Cohen, Vitalo, Anthony, & Pierce, 1980; Cohen, Danley, & Nemec, 1985). Similarly, some settings strongly emphasize support development (Beard et al., 1982) whereas other settings focus more on the development of client skills (Azrin & Philip, 1979).

9. Long-term drug treatment is an often necessary but rarely sufficient component of a rehabilitation intervention. In the last several decades, drug treatment has been provided to almost everyone who needs long-term psychiatric treatment. Ayd (1974) reported that up to 90% of hospitalized chronic patients are prescribed at least one neuroleptic drug. Dion, Dellario, and

Farkas (1982) found that 96% of a sample of severely disabled inpatients and outpatients in a Massachusetts catchment area were taking at least one neuroleptic drug. Similarly, Matthews, Roper, Mosher, and Menn (1979) reported that 100% of young, first-admission schizophrenics treated at the inpatient unit of a community mental health center received drug treatment. In general, it appears that 90–100% of psychotic patients who are hospitalized for treatment receive neuroleptic drug treatment.

Drug treatment is universally applied despite the evidence that some people do not need or want it and that some innovative treatment programs have achieved as good or better treatment success without drug treatment. In 1976, Gardos and Cole concluded that 50% of patients on outpatient maintenance medication may not need to be on maintenance medication. A large number of studies involving neuroleptics report similar findings. Usually, from 20- 50% of those on neuroleptic drugs are reported to relapse, whereas from 20–30% of those on placebo fail to relapse. Findings such as these suggest that a substantial number of clients continue to receive neuroleptic drugs who can function as well without them. Many patients do not need neuroleptics either because they can do just as well without them or because an active rehabilitation program mitigates the need for drugs (Carpenter, Heinrichs, & Hanlon, 1987; Matthews et al., 1979; Paul, Tobias, & Holly, 1972). In addition to these intervention studies, major reviews of the literature have indicated that treatment with lower doses of neuroleptics can be equally as effective as much higher doses of the same medications. The Task Force on Tardive Dyskinesia (1979) has proposed the use of drug holidays (designated periods of time that are drug-free), on-off trials to test necessity for neuroleptics at regular intervals, and dosage-lowering trials in an attempt to decrease the amounts of neuroleptics to which patients are exposed (Kane, 1987). Intermittent neuroleptic medication of some patients, rather than maintenance medication is another approach currently being explored (Carpenter, McGlashan, & Strauss, 1977; Carpenter, Heinrichs, & Hanlon, 1987; Herz, Szymanski, & Simon, 1982).

Despite much research on neuroleptic therapy, little attention has been paid to its effects on rehabilitation outcome, that is, on the client's ability to function in specific environments. A review of maintenance neuroleptic therapy by Docherty, Sims, & van Kammen (1975) found only 4 of 31 studies that measured the effectiveness of neuroleptics on dimensions other than symptoms and readmissions. Englehardt and Rosen (1976) assert that drug treatment alone has been insufficient in addressing a person's ability to function in residential or vocational environments.

The lack of a demonstrated relationship between drug treatment and rehabilitation should come as no surprise. Drug treatment simply does not develop the person's skills, energy, and community supports necessary for living, learning, socializing, and working in the community. Although the effect of drug therapy on symptomatic behavior may be thought of as preparing the person for rehabilitation, the relationship between drug therapy and rehabilitation may also be viewed in a slightly different way. A rehabilitation intervention could be conceived of as supportive to the withdrawal of drug therapy. That is, once drug therapy has prepared the person for rehabilitation, a successful rehabilitation intervention might prepare the person for the reduction of drug therapy.

Another way in which rehabilitation may be supportive of drug therapy is to increase the probability of drug therapy compliance (Weiden, Shaw, & Mann, 1986). A review of noncompliance data indicates an average noncompliance of 48% for phenothiazine, 49% for antianxiety or antidepressant drugs, and 32% for lithium (Barofsky & Connelly, 1983). Following a drug regimen may be considered a skill that can be taught to a person who is psychiatrically disabled. Providers of drug therapy might also wish to consider whether a rehabilitation diagnosis provides any additional information relevant to the need for drug therapy. It could be that a client's skill level and supports may partially predict the client's response to chemotherapy.

In summary, drug therapy can be considered as an often necessary but rarely sufficient component of rehabilitation. From a rehabilitation perspective, drug therapy is a useful intervention but rarely an entire rehabilitation program. Likewise, practitioners of drug therapy may view rehabilitation as supportive in increasing drug compliance, determining the initial need for drug therapy, and/or decreasing drug dosage.

Summary

The psychiatric rehabilitation philosophy is the foundation of the psychiatric rehabilitation process and psychiatric rehabilitation technology. The model and principles firmly anchor both the process and the technology in a set of values and beliefs. In the next chapter the psychiatric rehabilitation process and technology are described.

5

Process and Technology

Integrity without knowledge is weak and useless, and knowledge without integrity is dangerous and dreadful.

Samuel Johnson

*T*he mental health field is prepared to accept a psychiatric rehabilitation approach as a preferred method for helping persons with severe psychiatric disabilities. The myths of the past have been discarded, outcome studies have begun to emerge, and the philosophy of psychiatric rehabilitation has been articulated. Unfortunately, however, a rehabilitation approach will not be adopted if the technology for implementing the rehabilitation process is not developed and used.

This chapter focuses on the technology used in the process

Parts of this chapter are excerpted with permission from:

Anthony, W. A., Cohen, M. R., & Farkas, M. (1987). Training and technical assistance in rehabilitation. In A. Meyerson & T. Fine (Eds.), *Psychiatric disability: Clinical, legal, and administrative dimensions* (pp. 251–269). Washington, DC: American Psychiatric Press.

Anthony, W. A., & Farkas, M. D. (1989). The future of psychiatric rehabilitation. In M. D. Farkas & W. A. Anthony (Eds.), *Psychiatric rehabilitation programs: Putting theory into practice* (pp. 226–239). Baltimore: Johns Hopkins University Press.

of psychiatric rehabilitation. The emerging consensus concerning the fundamental phases of the psychiatric rehabilitation process is presented, followed by a description of technology a practitioner uses to help a person with a psychiatric disability participate in and benefit from the rehabilitation process.

Psychiatric Rehabilitation Process

The philosophy of rehabilitation provides direction to the psychiatric rehabilitation process: that is, disabled people need skills and support to function in the residential, educational, social, and vocational environments of their choice. The psychiatric rehabilitation process consists of three phases designed around the development of wanted and needed skills and supports: the diagnostic phase, the planning phase, and the intervention phase.

The *diagnostic phase* involves the practitioner helping the client to evaluate his or her skill and support strengths and deficits. In contrast to the traditional psychiatric diagnosis that describes symptomatology, the rehabilitation diagnosis yields a behavioral description of the person's current skill functioning and an operational definition of the current level of environmental support in the chosen residential, educational, social, and/or vocational environments. The diagnostic information enables the rehabilitation practitioner to help the client to develop a rehabilitation plan in the *planning phase*. A rehabilitation plan specifies how to develop the person's skills and/or supports to achieve the person's rehabilitation goals. The plan is similar to an individualized service plan, the major difference being its identification of high-priority skill and resource development objectives and specific interventions for each objective, rather than identification of service providers. In the *intervention phase*, the rehabilitation plan is implemented to develop the person's skills and/or environment to be more supportive of the person's functioning.

Rehabilitation Diagnosis

Practitioners conducting a rehabilitation diagnosis first involve clients in choosing the environments in which they intend to function in the next 6–18 months, (i.e., setting the clients' overall rehabilitation goals). For example, John Grace and his practitioner may agree that John wants to attend the After Hours Club until October of next year. This overall rehabilitation goal is the basis for a subsequent skill and support assessment.

By means of a functional assessment, the practitioner and client develop an understanding of those skills the client can and cannot perform related to achieving the client's overall rehabilitation goal. Practitioners work with clients to list and describe the unique skills the client needs based on the behavioral requirements of the chosen environment, as well as the behaviors personally important to the client's satisfaction in the chosen environment. For example, one skill John needs to use at the After Hours Club is demonstrating understanding. John and his practitioner describe the use of this skill as the percentage of times per week John restates feelings and the reasons for those feelings when conversing with club members during social events. Once use of the skill is described, John and his practitioner evaluate John's present and needed use of this skill at the After Hours Club.

A resource assessment evaluates the presence or absence of supports critical to the client's achieving the overall rehabilitation goal. By means of a resource assessment, the practitioner involves the client in listing and describing those persons, places, things, and activities necessary for the client to be successful in the chosen environment. Again, the listed resources are based on both environmental requirements and personal needs. For example, the After Hours Club may require its new members to have an identified member as a buddy. John and his practitioner describe this needed resource as the number of times during the weekend a designated member telephones John, when John is at home during the weekend. Once the resource is described, John and his practitioner evaluate the present and needed level of support provided by the resource. Tables 5–1 and 5–2 present an example of a portion of a functional assessment and a resource assessment from John's records.

Rehabilitation Planning

Based on the diagnosis, high-priority skill and resource development objectives are identified. The practitioner assigns a

TABLE 5–1 *Example: Functional Assessment Chart*

Overall Rehabilitation Goal: John attends the After Hours Club until October.

Strengths/ Deficits	Critical Skills	Skill Use Descriptions	Skill Evaluations[1]					
			Spontaneous Use		Prompted Use		Performance	
			Present	Needed	Yes	No	Yes	No
+	Suggesting activities	The percentage of times per week that John presents an idea for an activity that he and another member could do when discussing plans for the following day with a club member.	50%	50%				
–	Demonstrating understanding	The percentage of times per week John restates feelings and the reasons for those feelings when conversing with other club members during social events.	0	70%		X		X
–	Offering support	Percentage of times per month John makes a comforting statement to a member who has expressed distress.	0	90%	X		X	

[1] The client's skill level is evaluated in the three different ways. The spontaneous use column indicates the client's highest present level of spontaneous use of the skill in the target environment as compared to the needed level of skill use. The prompted use column indicates whether the client can (Yes) or cannot (No) perform the skill at least once in the target environment. The performance column indicates whether the client can (Yes) or cannot (No) perform the skill in an assessment or learning environment. If the client's present level of spontaneous skill use is zero, then prompted use is evaluated. Similarly, if the client has been evaluated as unable to apply (No) the skill when prompted, then skill performance is evaluated.

TABLE 5–2 Example: Resource Assessment Chart

Overall Rehabilitation Goal: John attends the After Hours Club until October.

Strengths/ Deficits	Critical Resources	Resource Use Descriptions	Present	Needed
–	Member contacts	Number of times per week a designated club member telephones John when he is home for the weekend.	0	2
+	Spending money	Number of times per month John's parents give him $100.00 for his social expenses.	4	4
–	Organized activities	Number of times per week the clubhouse program provides at least 2 hours of structured activities during the day for John.	4	5

specific intervention for each skill or resource objective in the plan and organizes the responsibilities for providing the intervention selected for each objective. The client and practitioner sign the rehabilitation plan to indicate their agreement. Table 5–3 presents an example of a portion of a rehabilitation plan from John's records.

Rehabilitation Interventions

The rehabilitation practitioner uses two major types of interventions: skill development and resource development. These interventions improve the client's use of skills or supports. There are two ways to develop skills. The first, direct skills teaching, is used when the functional asessment indicates the client has not acquired the skill (i.e., the client cannot perform the skill in the assessment situation). Direct skills teaching leads the client through a systematic series of instructional activities resulting in the client's competent use of new behaviors (Cohen, Danley, & Nemec, 1985). Direct skills teaching is unique in its use of comprehensive teaching methods that result in the client's learning to use the skill when needed.

The second way to develop skills, skills programming, prescribes a step-by-step procedure to prepare the client to use existing skills as needed (Cohen, Danley, & Nemec, 1985). Skills programming is designed to help clients overcome barriers to their regular use of a skill they already can perform. Skills programming prepares the client to use a skill as frequently as needed in a particular environment.

Direct skills teaching and skills programming complement one another. For example, after using direct skills teaching to enable John to perform the skill of demonstrating understanding in simulated situations during teaching sessions, the practitioner, together with John, would then use skills programming to generate a series of activities to overcome the barriers that prevent him from using this skill outside the teaching sessions. For instance, the practitioner and John would devise steps to help John overcome a barrier such as forgetting to use the demonstrating understanding skill in the After Hours Club as frequently as needed.

In contrast to skill development interventions, resource development interventions are designed to link the client with a

TABLE 5–3 Example: Rehabilitation Plan

Overall Rehabilitation Goal: John attends the After Hours Club until October.

Priority Skill/Resource Development Objectives	Interventions	Person(s) Responsible	Starting Dates	Completion Dates
John demonstrates understanding 70% of the times/week when conversing with other club members during social events.	Direct skills teaching Skills programming	Occupational therapist Clubhouse staff	Jan. 1 April 16	April 15 June 15
Clubhouse program provides organized activities 5 times per week.	Resource modification	Clubhouse director	Feb. 1	April 15

I participated in developing this plan and the plan reflects my objectives. Client's signature: _____

PSYCHIATRIC REHABILITATION

resource that presently exists (resource coordination) or to modify resources that do not function in the particular way needed by the client (resource modification). Resource coordination involves selecting a preferred resource, arranging for its use, and supporting the client in following through on using the resources. For example, in order for John to attend the After Hours Club he needs transportation. In this example, John and his practitioner would work to clarify the important values in making a choice about the provider of the transportation (e.g., cost, availability, reliability). Once the resource is selected, John is assisted in linking with the resource. The practitioner works with both the provider of the transportation (e.g., Share-A-Ride) and John to overcome any barriers that might prevent John from successfully using the transportation resource.

Resource modification is the technique of adapting an existing resource to fit the needs of the client better. For example, in order to provide a structured activity for John 5 days a week, the practitioner and John might arrange a shared volunteer position at another social service program. The skills of resource modification involve negotiating with community resources or supportive persons to make changes to meet a particular client's resource needs. Resource coordination and resource modification are similar to the practice of case management (see chapter 10). The difference is that resource coordination and resource modification are the major interventions of a classic case management intervention, while in rehabilitation they are only part of a more comprehensive rehabilitation approach.

In summary, the rehabilitation process unfolds as rehabilitation practitioners engage clients and significant others in diagnosing, planning, and developing the skills and supports required by clients to be successful and satisfied in a particular environment. The practitioner's ability to involve clients in the rehabilitation process is facilitated by the practitioner's level of skills.

Psychiatric Rehabilitation Technology

The psychiatric rehabilitation process of diagnosing, planning, and intervening can be facilitated by practitioners expert in the technology of psychiatric rehabilitation. A technology can be defined as the application of scientific knowledge to the solution of individual or societal problems, and/or the attainment of individual or societal goals. In the case of psychiatric rehabilitation, the technology can be thought of as a human technology (Carkhuff & Berenson, 1976) as differentiated from a mechanical technology. That is, a human technology applies scientific knowledge to achieve human resource development goals rather than industrial or commercial goals. In psychiatric rehabilitation, the goals are related to reducing the disability and handicap for persons with long-term psychiatric problems.

Scientific Knowledge

The scientific basis of psychiatric rehabilitation technology is the research literature referenced throughout this text. Researchers from a variety of mental health and rehabilitation disciplines have studied and evaluated the rehabilitation process. The technology of psychiatric rehabilitation practice is based as much as possible on what has been learned to date by means of the scientific method.

The scientific method or approach to knowledge development provides "an objective set of rules for gathering, evaluating, and reporting information" (Cozby, 1989, p. 5). "There are built-in checks all along the way to scientific knowledge. These checks are so conceived and used that they control and verify the scientist's activities and conclusions to the end of attaining dependable knowledge outside himself" (Kerlinger, 1964, p. 7). Basing the psychiatric rehabilitation technology on what has been learned from the scientific method provides the practitioner with a more credible and viable technology.

The purposes of the scientific method are essentially to make discoveries, to learn facts, and to advance knowledge (Kerlinger, 1964). All types of research methodologies are based on the scientific method. Research methodologies in psychiatric rehabilitation are, by necessity, diverse because of the diverse research questions. As a result, psychiatric rehabilitation technology has been developed from what has been learned from data gathered by a variety of methodologies, including methods variously referred to as experimental and quasi-experimental research, survey research, clinical research, experimental single-subject designs, evaluation research, program evaluation research, and exploratory data analysis. All of these research strategies represent valid ways of creating new knowledge from which technology can be developed. In the hierarchy of research methods, experimental research studies have the greatest degree of rigor. However, less rigorous methods have up to now been the greatest source of knowledge from which psychiatric rehabilitation technology has been derived.

Technology of Psychiatric Rehabilitation Practice

The psychiatric rehabilitation technology operationally defines the knowledge and skills the practitioner uses to assist clients through the rehabilitation process. The essential components of psychiatric rehabilitation technology are clearly defined practitioner skills and the underlying knowledge about how to use the skills most effectively.

The technology of psychiatric rehabilitation is based on an understanding of the skills the practitioner needs to help clients participate in the psychiatric rehabilitation process of diagnosing/planning/intervening. Table 5–4 lists the activities of practitioners during the three phases of the psychiatric rehabilitation process.

The Center for Psychiatric Rehabilitation so far has identified and operationally defined more than 70 practitioner skills useful in helping clients to participate in the psychiatric rehabilitation process. These skills and the knowledge necessary to implement these skills effectively are combined in unique ways to form the technology of psychiatric rehabilitation. In order to guide the client through each phase of the rehabilitation process, the practitioner needs to be expert in the technology of setting an overall rehabilitation goal, functional assessment, resource assessment, rehabilitation planning, direct skills teaching, skills programming, resource coordination, and resource modification. Table 5–5 presents an example of the technology for setting an overall rehabilitation goal.

The Center for Psychiatric Rehabilitation has packaged much of this technology into multimedia training packages capable of improving a practitioner's ability to learn and apply the various skills that make up this technology (M. R. Cohen et al., 1985,

TABLE 5–4 Activities of Practitioners during the Three Phases of the Psychiatric Rehabilitation Process

Phases:	Diagnosing	Planning	Intervening
Activities:	*Setting an overall rehabilitation goal* • Assessing rehabilitation readiness • Connecting with clients • Identifying personal criteria • Describing alternative environments • Choosing the goal	*Planning for skills development* • Setting priorities • Defining objectives • Choosing interventions • Formulating the plan	*Direct skills teaching* • Outlining skill content • Planning the lesson • Programming skill use • Coaching the client
	Functional assessment • Listing critical skills • Describing skill use • Evaluating skill functioning • Coaching the client	*Planning for resource development* • Setting priorities • Defining objectives • Choosing interventions • Formulating the plan	*Skills programming* • Identifying barriers • Developing the program • Supporting client action
	Resource assessment • Listing critical resources • Describing resource use • Evaluating resource use • Coaching the client		*Resource coordination* • Marketing clients to resources • Problem solving • Programming resource use
			Resource modification • Assessing readiness for change • Evaluating resources • Proposing change • Consulting to resources • Training resources

TABLE 5-5 Example: Technology for Setting an Overall Rehabilitation Goal

Activities:	Assessing Rehabilitation Readiness	Connecting with Clients	Identifying Personal Criteria	Describing Alternative Environments	Choosing the Goal
Skills:	• Inferring perceived need	• Orienting	• Clarifying values	• Specifying alternative environments	• Defining criteria
	• Validating commitment to change	• Demonstrating understanding	• Analyzing experiences	• Defining critical characteristics	• Evaluating alternative environments
	• Estimating awareness	• Self-disclosing	• Inferring personal criteria	• Researching alternative environments	• Specifying the goal
	• Discriminating closeness needs	• Inspiring			
	• Judging readiness				

From: Cohen, M., Farkas, M., Cohen, B., Unger, K. (1990): Psychiatric rehabilitation technology: Setting an overall rehabilitation goal. Center for Psychiatric Rehabilitation, Boston University, Boston, MA.

1986, 1990). Research during the last decade has documented the increased skill level of practitioners exposed to this technology (Farkas & Anthony, 1989; Farkas, O'Brien, & Nemec, 1988; Goering, Wasylenki, et al., 1988; National Institute of Handicapped Research, 1980; Rogers, Cohen, Danley, Hutchinson, & Anthony, 1986).

In addition to the technology developed by the Center for Psychiatric Rehabilitation, other technological developments are relevant to the psychiatric rehabilitation approach. Unlike the technology developed by the Center for Psychiatric Rehabilitation, this technology was not designed to facilitate a client's progress through the psychiatric rehabilitation process, but it can be adapted for that purpose. Examples of this technology include the technology of social skills training (Liberman, Mueser, & Wallace, 1986) and human resource development (Carkhuff & Berenson, 1976).

Another type of technology that may be useful for psychiatric rehabilitation practitioners is case management technology (see chapter 10). Once again, although this technology was not developed specifically to facilitate the psychiatric rehabilitation process, it can be useful to psychiatric rehabilitation practitioners (Goering, Wasylenki, et al., 1988).

All practitioner-level technology, whether or not it is actually designed to facilitate a client's progress through the psychiatric rehabilitation process, can help practitioners *do* their jobs more effectively. Changes in practitioner performance are the obvious results of learning a technology. However, these improved skills are based on current knowledge that guides the use of these skills. A comprehensive psychiatric rehabilitation technology also includes the knowledge needed to perform the skills at the proper time, and to overcome the barriers interfering with the use of the skills.

At first glance, the psychiatric rehabilitation process of diagnosing/planning/intervening seems so elementary and logical that one might think no special technology is needed. Unfortunately, that is not the case. As an analogue, consider the experimental research process, a fairly straightforward process for investigating possible cause/effect relationships by exposing one or more experimental groups to one or more treatment conditions and comparing the results to one or more control groups not receiving the treatment. Most grade school students can understand the experimental research process. Yet conducting experimental research in a given field demands a certain technology without which the process becomes sloppy and ineffective. Even though the process can occur with poor technology, the possibility of useful outcomes is also poor. So it is with the psychiatric rehabilitation process.

Resistance to Psychiatric Rehabilitation Technology

Some people resist adoption of psychiatric rehabilitation technology because they view the technology as a mechanical rather than human technology. They do not appreciate the interpersonal components of the technology that are at the very core of the technology, insuring that the technology is humanistic. They view psychiatric rehabilitation the way many people regard medical practice. Medical educators do not teach the human aspects of medical practice with the same verve and expertise as they do the medical technology. As a result, the human relationship is often missing in health care delivery. In contrast, the psychiatric rehabilitation technology emphasizes connecting with clients and is grounded in a person-oriented philosophy.

Other people resist the adoption of psychiatric rehabilitation technology because they do not value technology. The antitechnology forces believe that if practitioners' values are appropriate and if they work in a correctly structured rehabilitation program, then they will be able to help persons with psychiatric disabilities. Although this may sometimes be true, the question is, Can these practitioners be even more helpful if they are equipped with a psychiatric rehabilitation technology? Can the residential, educational, and vocational outcomes of persons with psychiatric disabilities be further improved? Using a medical analogy again, can we progress from the 19th-century doctor whose values were in the right place but whose technology was extremely limited, to a modern 21st-century physician, steeped in technology and whose values are still in the right place? The 21st-century practitioner can be grounded in the humane principles of psychiatric rehabilitation and equipped with a set of techniques and tools capable of helping persons with psychiatric disabilities benefit from the rehabilitation process.

Psychiatric Rehabilitation Technology
and the Mental Health Culture

The adoption of psychiatric rehabilitation technology is a gradual process. Practitioners and programs progress through a series of evolutionary steps in their effort to incorporate the skills and techniques of psychiatric rehabilitation into their practice and programs. This step-by-step evolutionary process is how new technology is normally incorporated into any field. Technology is rarely adopted suddenly; more often the course is steady, year-to-year modifications.

However, even this slow, steady adoption is uncertain if the mental health culture is not supportive of a psychiatric rehabilitation approach. The mental health culture is the values and norms of the persons in the mental health field. Fortunately, the values of such persons as policy makers, administrators, consumers, and practitioners are becoming more consistent with the psychiatric rehabilitation philosophy.

A culture has a strong influence on the adoption of any technology. Acceptance of the technology is not necessarily due to the worth of the innovation itself, but rather to the readiness of the culture to accept it. The best example outside the mental health field is the differential adoption of western technology by Japan and China. Major reasons for the difference between Japan's and China's acceptance of western technology are their cultural differences and the cultural influence on the adoption of technology.

As discussed by Anthony and Farkas (1989), several changes in the mental health culture are leading to greater acceptance of psychiatric rehabilitation technology. One major change is a shift in defining legitimate outcome goals for state and local departments of mental health. For example, residential and vocational goals are now becoming acceptable mental health goals. Mental health professionals are increasingly assisting the vocational and residential functioning of persons with psychiatric disabilities. Policy makers are beginning to understand that vocational and residential functioning are not just the concern of the vocational rehabilitation or welfare departments or of the housing authority, but are also the proper domain of mental health professionals. Increasing numbers of state mental health directors are concerned about the vocational and residential functioning of persons with severe psychiatric disabilities. That is a big change from a decade ago. Psychiatric rehabilitation technology can grow in this new culture.

Another relevant change is that the mental health culture is beginning to see persons with psychiatric disabilities as persons first and disabled persons second. That cultural change happened years ago in the area of physical disabilities when persons with physical disabilities said, "Wait a minute, I'm not just a 'wheelchair person,' I'm a *person* who uses a wheelchair." Suddenly, consumers of mental health services are saying, "I'm not just mentally ill. I'm a *person* who has a mental illness, or some psychiatric problems, or some emotional problems." The *person* is being emphasized. The person with a disability is somebody who has aspirations just like anyone else, for a job, for a place to live, for friends, for people to turn to in a crisis. The field is learning that persons with psychiatric disabilities are not disabled 24 hours a day; at many times they are not disabled at all. They are people first. This change in thinking is consistent with the philosophy and technology of psychiatric rehabilitation.

Another change in the mental health culture is that the environment is now being recognized as a major factor in a person's recovery. No longer is the person assessed and treated in a vacuum, as though simply changing the client's strengths and deficits will result in the client's recovery. The mental health culture now realizes that both the person and the environment must be assessed—the immediate as well as the larger environment (for example, the person's home as well as the social security system or the public welfare system). These relevant systems can retard or prevent the person's rehabilitation. When the mental health culture sees the environment as a barrier—and as a potential facilitator—the mental health culture becomes open to a rehabilitation philosophy and technology that address environmental barriers and opportunities.

The mental health culture is also recognizing the importance of community supports. In physical rehabilitation, supports have always been seen as critical. Physical rehabilitation supports are concrete—crutches and wheelchairs, canes, ramps, and so on. In psychiatric rehabilitation, whether the goals are vocational, independent living, or educational, supports are also critical. More frequently today, mental health professionals ask, What supports are being provided? Programs such as supported employment, supported housing, supported education, and consumer-run alternatives provide peer supports, companions, professional supports, and/or support networks. Support has become an accepted type of intervention. Much remains to be learned about how to provide support, but the provision of support is becoming a routine intervention in the mental health culture. This change bodes well for the future adoption of psychiatric rehabilitation technology.

Currently, consumers and family members are demanding that mental health professionals describe the helping process in more understandable ways. They are not as awe struck as they once were of professionals who make treatment sound mysterious and esoteric. Fortunately, the practice of psychiatric rehabilitation and its accompanying technology can be explained in a way that consumers and their families can understand. As the mental health culture increasingly uses more comprehensible language, it is becoming more consistent with psychiatric rehabilitation philosophy and technology. Informed consent is an example of how the helping process can be described in a more straightforward manner.

In order for psychiatric rehabilitation to be most effective, the mental health culture must see it as human and not mechanical, as based on scientific facts and not anecdotal impressions. A mental health culture that values both humanness and effectiveness will be beneficial to the adoption of psychiatric rehabilitation technology.

Psychiatric rehabilitation practitioners differ in how committed they are to using psychiatric rehabilitation technology to improve their practice. However, as consensus continues to build on the basic philosophy and outcomes of psychiatric rehabilitation, as the process of psychiatric rehabilitation becomes even better understood, and as the mental health culture becomes more accepting of a human technology, more practitioners and administrators will see psychiatric rehabilitation technology as a means for acting more forcefully on their beliefs, for better involving their clients in the process, and most important, for achieving improved rehabilitation outcomes for their clients.

Summary

At its most basic level, the process of psychiatric rehabilitation seeks to help persons with psychiatric disabilities determine their goals, identify what they need to do and to have in order to achieve these goals, plan what to work on first and how, and then develop the necessary skills and/or supports to achieve their goals. Every psychiatric rehabilitation practitioner and setting engages clients in some or all of this process. Whether it is a clubhouse or a hospital, a day treatment program or a supported work program, a community residence or a community college, the process of psychiatric rehabilitation can be facilitated by the adoption of technology. When the technology is used, the impact of psychiatric rehabilitation can be assessed. The next two chapters provide further description and discussion of psychiatric rehabilitation diagnoses, plans, and interventions.

6
Diagnoses

> *It must be borne in mind that the tragedy of life doesn't lie in not reaching your goal. The tragedy lies in having no goal to reach.*
>
> *Benjamin E. Mayes*

Psychiatric rehabilitation diagnoses help persons with psychiatric disabilities identify their overall rehabilitation goals and what they must do and have in order to achieve their goals. A psychiatric rehabilitation diagnosis must not be confused with a traditional psychiatric diagnosis. The goal, process, and tools of each diagnosis are different. Yet, each diagnosis provides useful and meaningful information, requires a trained diagnostician, and has a role in a comprehensive treatment and rehabilitation intervention.

The focus of this chapter is on psychiatric rehabilitation diagnosis. However, in order to highlight the differences between rehabilitation diagnosis and the traditional, better-known psychiat-

Parts of this chapter are excerpted with permission from:

Anthony, W. A., Cohen, M. R., & Nemec, P. (1987). Assessment in psychiatric rehabilitation. In B. Bolton (Ed.), *Handbook of measurement and evaluation in rehabilitation* (pp. 299–312). Baltimore: Paul Brookes.

Cohen, B. F., & Anthony, W. A. (1984). Functional assessment in psychiatric rehabilitation. In A. Halpern & M. Fuhrer (Eds.), *Functional assessment in rehabilitation* (pp. 79–100). Baltimore: Paul Brookes.

ric diagnosis, an example of each is provided. The contrast between these two approaches illustrates the unique contributions of a psychiatric rehabilitation diagnosis.

A Case Example of a Traditional Psychiatric Diagnosis and a Psychiatric Rehabilitation Diagnosis

The following excerpts are from the case file of a person with a psychiatric disability. The same client has both a psychiatric diagnosis and a rehabilitation diagnosis.

The client, Richard, a 25-year-old, single, white male was referred to a clinic for evaluation. He expressed feelings of sadness and difficulty motivating himself at work. He was employed full-time as a stock clerk. Living at home with his parents, Richard also attended the evening program of the Day Treatment Center. Approximately 1 month before his referral to the clinic, Richard's work supervisor told him that his work productivity was declining. Around this same time, Richard began spending more time in bed on weekends and was told by the Day Treatment Center staff that he seemed lethargic.

Richard had completed high school and attended community college part-time, majoring in accounting and business. After completing the college program and obtaining his first job, he experienced his first psychotic break. He quit that job and was employed irregularly as a day laborer and a stock boy.

Richard's psychiatric treatment began at the age of 22. He was seen privately by a psychiatrist for delusions and stabilized with haldol (5 mg.qd.). A few years later he began attending the day treatment program at the local community mental health center. After 6 months, Richard obtained his current job with the assistance of the state Division of Vocational Rehabilitation.

Traditional Psychiatric Diagnosis (Excerpts from the File)

Richard is oriented by person, place, and time. He presents a neat and appropriate appearance. His behavior is socially acceptable. He appears to be slightly above average in intelligence, although psychological testing places him within the average range.

94

Some impairment in thought process is evident—some idiosyncratic word usage, difficulty in concentration, concreteness, and overattention to detail. He appears somewhat anxious and suspicious as well. Richard denies any current delusions, hallucinations, suicidal or homicidal ideation. A thought disorder and an affective disorder are clearly both present. Considering his psychiatric history and response to medication, the thought disorder is assumed to be primary.

Affective disturbance seems limited to depression, possibly in response to deterioration from premorbid functioning. No evidence of mania or hypomania is apparent at this time. In view of his dependence on others, tendency toward rebelliousness, and an inclination to procrastinate, passive aggressiveness is suggested.

DSM III

Axis I: 295.62 Schizophrenia, residual type, chronic
           296.82 Atypical depression
Axis II: 301.84 Passive-aggressive personality disorder
           (provisional)
Axis III: No information
Axis IV: Psychosocial stressors: new job, new girl friend
           Severity: 4—Moderate
Axis V: Highest level of adaptive functioning in past year
           4—Fair

Psychiatric Rehabilitation Assessment (Excerpts from the File)

Richard's overall rehabilitation goal is to continue at his present full-time job as an inventory clerk for at least the next year. He reports some need for improvement in his job performance, in agreement with his supervisor who indicates that Richard's work productivity has declined.

Based on discussion and self-evaluation at work, the following skill and resource strengths and deficits have been identified.

Skills

Organizing work tasks: Number of days per week Richard lists in order of priority the work tasks to be completed that day.
Present level: 0   Needed level: 5
(Skill deficit)

95

Requesting assignments:	Number of days per week Richard asks his supervisor in the morning which tasks he should complete that day. Present level: 3 Needed level: 5 (Skill deficit)
Stocking supplies:	Percentage of times per day Richard piles the supplies that have been requisitioned on the loading platform in the morning before the end of the workday. Present level: 100% Needed level: 100% (Skill strength)
Preparing inventory:	Number of days per week Richard lists the number and type of items left on the shelves before leaving work. Present level: 3 Needed level: 5 (Skill deficit)
Conversing about impersonal topics:	Percentage of times per day Richard talks about subjects other than himself during informal conversations with his co-workers. Present level: 0% Needed level: 75% (Skill deficit)

Resources

Transportation:	Number of days per week someone drives Richard to and from work in the carpool. Present level: 5 Needed level: 5 (Resource strength)
Social environment:	Number of days per week the Day Treatment program (or alternative social setting) is open for Richard to drop in after work. Present level: 1 Needed level: 2 (Resource deficit)
Work counselor:	Number of days per month Margaret provides support and skill training to Richard around on-the-job problems. Present level: 2 Needed level: 4 (Resource deficit)

As the case example demonstrates, a rehabilitation diagnosis and a psychiatric diagnosis focus on completely different aspects of the person. In contrast to the traditional diagnostic focus on pathology and symptom development over time, the rehabilitation diagnosis focuses on the skills and the resources the person needs to achieve an overall rehabilitation goal. Rather than assigning particular diagnostic categories, the rehabilitation diagnosis describes the person's skills and resources with respect to the client's attainment of the overall goal—which in Richard's case is to continue to work full-time as an inventory clerk for another year.

The goal of the traditional psychiatric diagnostic procedure is to assign a diagnostic label to describe the client's pathological symptoms based on the client's history, signs, and symptoms. In contrast, the goal of the rehabilitation assessment is to describe the client's skills and the environmental resources with respect to their impact on the client's overall rehabilitation goals based on the client's and significant others' perspectives and objective evaluation.

Because the goals of the two approaches are so different, it would be expected that the diagnostic procedures would be different. Just as psychiatric knowledge and specific diagnostic techniques are needed by a practitioner to conduct a psychiatric diagnosis, a practitioner also needs unique knowledge and techniques to conduct a psychiatric rehabilitation diagnosis. This psychiatric rehabilitation diagnostic technology must be mastered if a practitioner is to become expert in conducting psychiatric rehabilitation diagnoses.

The Empirical Foundation for a Psychiatric Rehabilitation Diagnostic Approach

Recent developments in psychiatric rehabilitation diagnosis reported in this chapter are anchored in empirical studies conducted during the last several decades by many researchers from a variety of disciplines. In a series of reviews of this research literature, Anthony and others have concluded that rehabilitation outcome of clients with psychiatric disabilities is a function of clients' skills and supportive resources in the community.

In the initial review of this body of research, Anthony and Margules (1974) concluded that persons with long-term psychiatric disabilities can learn a variety of skills regardless of their symptomatology and that these skills, when properly integrated into a comprehensive rehabilitation program that provides support for the use of these skills in the community, can have a significant impact on client rehabilitation outcome. Since that 1974 literature review, several other reviews have also concluded that rehabilitation outcome is a function of clients' skills and the supportive resources in their communities (Anthony, 1980; Anthony, Cohen, & Vitalo, 1978; Anthony, Cohen, & Cohen, 1984; Anthony & Jansen, 1984; Cohen & Anthony, 1984; Dion & Anthony, 1987).

Given that rehabilitation outcome is a function of clients' skills and resources, it makes sense that the improvement of skills and resources be the focus of psychiatric rehabilitation interventions. *It follows logically that if rehabilitation interventions are designed to improve clients' skills and supports, then rehabilitation diagnoses evaluate clients' present and needed skills and supports.*

A technology for psychiatric rehabilitation diagnoses is needed because traditional psychiatric diagnostic procedures do *not* provide much information relevant to prescribing a rehabilitation intervention. Many of the reviews of the research literature report that little or no relationship exists among rehabilitation outcome, clients' psychiatric diagnostic labels, and descriptions of the clients' symptom patterns (Anthony, 1979; Anthony & Jansen, 1984; Cohen & Anthony, 1984). The lack of a relationship among psychiatric diagnosis, psychiatric symptoms, and rehabilitation outcome flows naturally from the lack of correlation between a person's symptomatology and functional skills. As discussed in chapter 2, measures of skills and measures of symptoms show little relationship to one another. For example, Townes and associates (1985) classified psychiatric patients into 6 groups, according to their unique pattern of strengths and deficits, and found that the classification was essentially independent of reported psychiatric symptomatology and diagnosis. Similar results were reported by Dellario, Goldfield, Farkas, and Cohen (1984). They correlated 16 different symptom measures with 19 different measures of function taken on the same psychiatric inpatients. Only 8 of the 304 correlations were statistically significant; no discernable pattern existed among these 8 correlations. With a correlation matrix of this size, 8 statistically significant correlations would be expected by chance.

In summary, the empirical literature suggests two conclusions. First, the present psychiatric diagnosis is neither descriptive, prescriptive, nor predictive with respect to rehabilitation. Thus, a unique diagnostic technology is needed for rehabilitation. Second, a psychiatric rehabilitation diagnosis needs to focus on describing clients' skills and environmental supports.

Components of a Psychiatric Rehabilitation Diagnosis

The psychiatric rehabilitation diagnosis is used to evaluate the client's skills and supports in the context of the environment in which the client chooses to live, learn, socialize, and work. The diagnosis contains three components: an overall rehabilitation goal, a functional assessment, and a resource assessment.

The overall rehabilitation goal identifies the particular environments in which the client chooses to live, learn, socialize, and work during the next 6 to 24 months (Cohen, Farkas, Cohen, & Unger, 1990). The particular environment may be one in which the client currently lives, learns, socializes, or works and wants to stay; or the environment may be one to which the client desires to move within the next year or two. The overall rehabilitation goal is established during a series of interviews with the client in which the client's personal criteria and alternative environments are explored. The overall rehabilitation goal is critical to the diagnosis because the hope of its achievement motivates the client to participate in the diagnosis.

Setting goals affects performance whether or not a person is disabled. A number of experimental studies have shown the positive effects of setting goals (Locke, Shaw, Saari, & Latham, 1981). "Goals affect performance by directing attention, mobilizing effort, increasing persistence and motivating strategy development" (Locke et al., 1981, p. 125). In addition, the overall rehabilitation goal focuses subsequent assessment of the client by limiting the skills and supports assessed to those that are relevant to satisfaction and success in that goal environment. The following are examples of overall rehabilitation goals:

- To live at Mulberry House until November 1992;
- To work in a sheltered workshop for one year;
- To study at a supported learning program at Worcester State College for the next two years.

99

The necessity of establishing the client's overall rehabilitation goal is consistent with the philosophy of psychiatric rehabilitation (Anthony, 1982; Cnaan et al., 1988). Taking the time to work with the client to set overall rehabilitation goals is also important because if this process is neglected, the practitioner and client very likely may be pursuing different goals without knowing it! Research evidence suggests that assessments of clients by practitioners and assessments by clients themselves often have little or no agreement on items as diverse as potential for recovery (Blackman, 1982), desired outcomes (Berzinz, Bednar, & Severy, 1975), rehabilitation issues (Leviton, 1973), perceptions of handicapping problems (Tichenor, Thomas, & Kravetz, 1975; Mitchell, Pyle, & Hatsukami, 1983), and the existence of functional skills (Dellario, Anthony, & Rogers, 1983). For example, Dimsdale, Klerman, and Shershow (1979) studied a group of hospital patients in which the staff viewed insight as the primary goal. The patients, however, placed insight at the bottom of their list of goals. Dimsdale and his colleagues concluded that if goals were shared by both practitioners and patients, patients might be more satisfied and the length of their hospitalizations might be reduced. Other research indicates that when clients' and therapists' goals are incongruent, clients do not appear to profit from therapy, are disappointed with their care, and often fail to comply with their treatment activities (Goin, Yamamoto, & Silverman, 1965; Lazare, Eisenthal, & Wasserman, 1975; Mitchell et al., 1983).

Sometimes the reasons for not involving clients in goal setting stems from the mistaken belief that consumers are unable to make decisions or choices. Some authors suggest that the inability of persons to make choices and set goals was related to their treatment environments. For example, Ryan (1976) suggests that the psychiatric treatment environment itself can take away a person's ability to make important life decisions and that the process of institutionalization results in a loss of initiative, an assumption of deviant values, and an inability to make decisions (Schmieding, 1968; Goffman, 1961).

Other researchers hold the view that lack of decision-making ability or goal setting is inherent in the pathology of mental illness. For example, the three major types of problems associated with schizophrenia include positive symptoms (e.g., hallucinations, delusions), negative symptoms (e.g., withdrawal, lack of goal-directed behavior and motivation), and disordered relationships (e.g. lack of personal ties). Of these three, negative symptoms are prognostically most important and are viewed as the source or result of chronicity (Keith & Matthews, 1984; Strauss, Carpenter, & Bartko, 1974; American Psychiatric Association, 1980). This view prevails even in the courts. Informed consent, or the right to choose treatments, can be legally denied persons with severe psychiatric disabilities, because they are judged to be incompetent to make such decisions (Schwartz & Blank, 1986).

Apart from disagreement as to whether persons with psychiatric disabilities can make their own choices or state their own needs, most people agree that setting goals in treatment is important (Carkhuff & Anthony, 1979). Some research evidence suggests that goal setting itself impacts outcome (Smith, 1976) and that the attainment of goals affects satisfaction and recidivism (Willer & Miller, 1978). Mental health professionals, however, still resist adopting goal setting as a regular part of their practice (Holroyd & Goldenberg, 1978) and, in particular, goal setting that reflects the consumer's perspective about desired rehabilitation outcomes (Farkas, Cohen, & Nemec, 1988). Thus, practitioners must become particularly skilled in helping persons with psychiatric disabilities set their overall rehabilitation goals.

After setting an overall rehabilitation goal, a practitioner can conduct a functional assessment, an evaluation of the client's functioning on those critical skills needed to achieve the overall rehabilitation goal. The functional assessment evaluates the frequency of the client's use of the critical skills described in such a way that the observable behavior and circumstances of skill use (e.g., what, where, when, with whom) are clearly stated (Cohen, Farkas, & Cohen, 1986). The functional assessment provides the practitioner and client with a complete picture of the client's skill strengths and deficits. Understanding the skill strengths increases confidence about the client's success in the specified environment. Understanding skill deficits points to the need for particular skill development interventions. The following is an example of a functional assessment of one critical skill—describing negative interactions.

Describing negative interactions is the percentage of times per week Joanne states the facts of what was said by whom, when the staff at the residence ask about her disagreement.
Present level of functioning: 20%
Needed level of functioning: 75%

Just as an assessment of the client's skills is critical for later skill development, a resource assessment is crucial to developing the client's supports. A resource assessment is an evaluation of the supports needed for the client to achieve the overall rehabilitation goal. The resource assessment evaluates the use of critical resources such as people, places, activities, and things. The critical resources are so described in the assessment that the observable characteristics and circumstances of use of the resource are clearly stated. An understanding of resource use clarifies the resources the client can rely on for success and satisfaction in the specified environment. The resource assessment points to the need for particular resource development interventions. The following is an example of a resource assessment of one critical resource—transportation.

Transportation is the number of days per week the van drives Charles roundtrip between home and the supported employment placement.
Present level of functioning: 3 days
Needed level of functioning: 5 days

The Psychiatric Rehabilitation Diagnostic Interview

The psychiatric rehabilitation diagnosis is formed in a series of interviews with the client. Two principles guide the diagnostic interview: (1) The practitioner attempts to maximize the involvement of the client in the interview; and (2) The information collected during the interview is recorded in a way that maximizes the clients' understanding of the assessment results.

Involving the Client

Involving the client in the interview means facilitating the client's active participation in formulating each part of the diagno-

102

sis. Client involvement increases the client's ownership of the rehabilitation diagnosis. The practitioner involves the client in the diagnostic interview by use of three skills: (1) orienting; (2) requesting information; and (3) demonstrating understanding (Cohen, Farkas, & Cohen, 1986).

Orienting. Orienting the client involves describing to the client the task, its goal, and the roles of both the practitioner and client (Cohen, Farkas, & Cohen, 1986). The orientation gives the client a clear picture of what will happen and how to participate. For example, the practitioner might say the following to orient the client during the beginning of a functional assessment:

Practitioner: The first task in functional assessment is listing critical skills. The goal of listing is to write a list of all the skills that you need to live successfully at home with your family. First, you and I will name the behaviors your family expects you to do. For example, they may want you to pay all your long-distance telephone charges each month. Second, you will tell me about the things you want to be able to do. For example, you may want to be able to do more organized activities with your friends. I will be asking you questions and summarizing what you say to make sure I'm understanding you. I want you to share your thoughts and feelings honestly and ask questions when you are unclear about something. Any questions?

Client: No.

Practitioner: Just to make sure you understand what we will be doing, please tell me, in your own words, what will happen next.

Client: Okay. We're going to discuss the behaviors my family wants me to do. I'm going to say what I want to do. I'm going to be honest with you.

The way the practitioner orients the client is important. The practitioner uses words the client is likely to understand and continually checks out the client's understanding of what the practitioner has previously said.

Requesting Information. Requesting information involves asking the client for facts, opinions, and feelings (Cohen, Farkas, & Cohen, 1986). Requesting information encourages the client to talk about a particular topic. Asking open-ended questions rather than direct questions encourages participation. For example, the practitioner might say the following to request information from the client during the listing of critical skills:

> Practitioner: Why do you think that communicating anger is a strength for you?

Demonstrating Understanding. Demonstrating understanding involves capturing in words what the client is feeling or thinking (Cohen, Farkas, & Cohen, 1986). Demonstrating understanding tells the client that the practitioner is listening and trying to gather the client's perspective. For example, the practitioner might say the following to demonstrate understanding to the client while listing critical skills:

> Client: I can't get my sister to hear that she's treating me bad. I try telling her how cruel she's being but she won't listen.
>
> Practitioner: You feel frustrated because she won't understand that she's hurting you.
>
> Client: Yeah, I keep trying to tell her but she just doesn't listen. Maybe she doesn't care.
>
> Practitioner: You're wondering if it doesn't matter to her if she hurts you.
>
> Client: Oh, I don't know. We were close as kids. Maybe it's me.

It is incumbent that practitioners demonstrate to their clients an understanding of their perspective. The research suggests only a moderate agreement between practitioner ratings and client ratings of the client's skills (Cook, 1983; Dellario, Anthony, & Rogers, 1983). The rehabilitation diagnostic interview includes strategies designed to check for and maximize client and practitioner agreement.

The three skills used to involve the client in the assessment —orienting, requesting information, and demonstrating understanding—are similar to interpersonal skills such as paraphrasing, responding to feelings, and question asking. Many clients do not know how to participate in an interview. They may be accustomed to psychiatric interviews that focus on their symptoms, maladaptive behaviors, and probable causes of impairment, with minimal attempts to involve the client in the interviews. Their inability to process new information efficiently may hamper their involvement (Öhman, Nordby, & d'Elia, 1986; Monti & Fingeret, 1987; Spaulding, Storms, Goodrich, & Sullivan, 1986). When the client does not participate in the interview, the practitioner must actively involve the client in the interview by using some or all of these three interpersonal skills.

Organizing the Information

The information gathered during a psychiatric rehabilitation diagnostic interview is usually recorded in a way that is consistent with agency record-keeping requirements. Records often vary in the amount of details recorded and the format of the record. Minimally, the final results of the diagnostic interview—the overall rehabilitation goal, a functional assessment, and a resource assessment—are recorded. Each of these three parts of the diagnosis has specific characteristics. The diagnosis recorded in the client's file, regardless of the particular agency's record-keeping format, should display the characteristics of clarity, brevity, and environmental specificity.

For example, an overall rehabilitation goal indicates a specific environment (or type of environment) and a date by which the goal should be achieved. Usually the overall rehabilitation goal is recorded in a simple, straightforward sentence at the top of functional assessment and resource assessment forms.

The functional assessment, in addition to being environmentally specific, clear, and brief, is skill-focused, behavioral, observable, and measurable. When recorded, the functional assessment names the critical skills, describes the needed use of these skills in a specific environment, and identifies the client's present and needed use of the skill. Table 6–1 is an example of how an agency might record the overall rehabilitation goal and the functional assessment on a form, reflecting these important record-keeping characteristics.

Finally, the resource assessment has many of the same record-keeping characteristics as the functional assessment, that is, observable, measurable, environmentally specific, clear, and brief. The resource assessment names the client's critical resources, describes the needed use of the resources, and identifies the client's present and needed use of these resources. Table 6–2 is an example of how an agency might record the resource assessment on a form reflecting the important record-keeping characteristics.

Psychiatric Rehabilitation Diagnostic Instruments

Literally hundreds of diagnostic instruments varying greatly in their focus exist for use with persons who are psychiatrically disabled. Some instruments focus on traditional psychiatric diagnosis and symptomatology, some on maladaptive behavior, some on level of functioning, and others on resources (both those needed by and/or available to the client). Many instruments contain items from several categories.

Several literature reviews have surveyed a variety of existing instruments potentially useful for psychiatric rehabilitation diagnosis. Anthony and Farkas (1982) described data-collecting strategies and instruments capable of measuring a broad range of possible client characteristics. The authors included a useful table of references that indicates the focus of the instrument, the person who completes it, the original population for whom it was developed, and information on reliability and validity.

From their review, Anthony, Cohen, and Nemec (1987) concluded that most existing instruments lack many characteristics that would be most useful for psychiatric rehabilitation diagnosis. Although increasing numbers of psychiatric diagnostic instruments are focusing on skill and resource assessments as opposed to symptoms and pathology, these instruments are still limited in their

TABLE 6-1 *Example: Functional Assessment Chart*

Overall Rehabilitation Goal: Betty wants to live at Nottingham House through February, 1992.

Strengths/ Deficits	Critical Skills	Skill Use Descriptions	Skill Evaluations[1]					
			Spontaneous Use		Prompted Use		Performance	
			Present	Needed	Yes	No	Yes	No
+	Dressing	Number of days/week Betty puts on matching clothes when going to the day treatment center.	3 days/ week	3 days/ week				
−	Following directions	Percentage of times/week Betty performs the steps of tasks as instructed, when her supervisor gives her an assignment.	0% of times/ week	90% of times/ week	X		X	
+	Speaking in turn	Percentage of times/week Betty says what she has to say after another person has finished speaking, when talking to the other residents during community meetings at the house.	95% of times/ week	85% of times/ week				
−	Refusing requests	Percentage of times/week Betty says "no" and the reason for her answer when other residents ask to borrow her things.	7 days/ week	5 days/ week		X		X

[1] The client's skill level is evaluated in three different ways. The spontaneous use column indicates the client's highest present level of spontaneous use of the skill in the target environment as compared to the needed (N) level of skill use. The prompted use column indicates whether the client can (Yes) or cannot (No) perform the skill at least once in the target environment. The performance column indicates whether the client can (Yes) or cannot (No) perform the skill in an assessment or learning environment. If the client's present level of spontaneous skill use is zero, then prompted use is evaluated. Similarly, if the client has been evaluated as unable to use the skill when prompted (No) then performance is evaluated.

TABLE 6-2   Example: Resource Assessment Chart

Overall Rehabilitation Goal: Betty wants to live at Nottingham House through February, 1992.

Strengths/ Deficits	Critical Resources	Resource Use Descriptions	Present	Needed
+	Parental support	Percentage of times/month parents visit Betty when asked by residence staff.	100% of times/ month	100% of times/ month
–	Social contact	Number of times/week friends and relatives tele-phone Betty when she is home on weekend pass.	0 times/week	2 times/week
–	Money	Number of dollars/week parents give Betty for household expenses.	$5.00/week	$25.00/week
–	Staff reinforcement	Number of times/week residence staff praise Betty when she does things she is supposed to do.	1 time/week	10 times/week

clinical application—most obviously by their lack of environmental specificity. Existing instruments, because they are standardized, provide information relevant to general environments rather than specific environments, for example, a general work setting rather than a specific job site.

Recently, some researchers have suggested that psychological and neuropsychological instruments be used with a rehabilitation perspective (Erickson & Binder, 1986). Rather than use these instruments to document psychiatric diagnosis, positive symptoms, and organicity, neuropsychological tests could be helpful in profiling a person's cognitive competences and deficits, for example, problem solving, concentration, reasoning, social judgment, information processing. This information could then be used in planning the rehabilitation intervention. Erickson and Binder (1986) have tentatively proposed a test protocol drawn from existing instruments. Rather than classify persons by psychiatric diagnosis, neuropsychological tests can identify unique competencies and deficits, independent of psychiatric diagnosis, that have obvious implications for the type of rehabilitation intervention (Townes et al., 1985).

The practitioner must use the information collected by such instruments in a conservative way because the information is but one source of data that leads toward the environmentally specific diagnosis illustrated in Tables 6–1 and 6–2. At the present state of instrument development, existing psychiatric rehabilitation diagnostic instruments are more valuable in research and program evaluation activities than in clinical practice. The many instruments now available that evaluate the overall functioning and resources of persons with psychiatric disabilities have proven more useful to researchers and evaluators than to practitioners (Anthony, Cohen, & Nemec, 1987).

In clinical diagnostic situations, the process itself must begin and end with the client. Before using any instruments, the practitioner attempts to obtain the client's perspective on the client's skill and resource strengths and deficits. The diagnosis can then proceed to acquiring information from significant others, testing the client's skill functioning, and observing the client's skill functioning in simulated environments.

Once all the data are gathered, the information is recorded

and reorganized in such a way that the client understands the diagnosis. Psychiatric rehabilitation diagnosis is not dominated by the instrument. It requires a practitioner able to develop a relationship with the client. Practitioners who conduct diagnostic interviews must have good interpersonal skills (e.g., the ability to demonstrate understanding). The practitioner must be skilled in involving clients in a psychiatric rehabilitation diagnostic process that the clients themselves understand.

Indeed, the focus and conduct of the diagnostic interview, rather than the diagnostic instruments, are the foundation for a valid diagnosis. Frey (1984, p. 35) has noted the limitations of assessment instruments: "Any attempt to capture, through single measures, an individual's status in a way that reflects all that is important to the rehabilitation process is ostentatious, to say the least." Or, as Spaulding et al. (1986, p. 574) have noted: "Patients' individual constellations of deficits are probably unique enough to preclude a 'cookbook' approach to treatment. Ideographic assessment and individualized treatment regimens will probably always be necessary."

Concluding Comment

Frey (1984) mentions that since the early 1940s rehabilitation (including psychiatric rehabilitation) has become a multidisciplinary endeavor. The field of psychiatric rehabilitation, conceived during the 1950s by the pioneering psychosocial rehabilitation centers (e.g., Fountain House in New York, Center Club in Boston, Horizon House in Philadelphia), did not really begin to evolve until the 1970s. Although the psychiatric rehabilitation field has considered itself multidisciplinary since its inception, nearly all of the early leaders were nonphysicians. The psychiatric diagnosis was in most instances considered irrelevant to rehabilitation, an assumption later confirmed by many empirical studies (Anthony, 1979). Yet rather than a new diagnostic technology being embraced as an alternative to traditional psychiatric diagnostic technology, most rehabilitation centers initially opted for a loosely structured, experiential assessment approach.

The adoption of the psychiatric rehabilitation diagnosis in

mental health and rehabilitation settings has been slowed by the mistaken assumption that a behaviorally specific assessment dictates a behavior modification intervention. In reality, the use of psychiatric rehabilitation diagnosis may have many advantages to a field composed of many disciplines (e.g., rehabilitation counselors, nurses, psychologists, social workers, occupational therapists, psychiatrists) who practice in many settings (e.g., psychosocial rehabilitation centers, state divisions of vocational rehabilitation, community mental health centers, workshops, state hospitals). A psychiatric rehabilitation diagnosis, grounded in a rehabilitation philosophy, can increase communication across disciplines, across settings, and with clients and their families. This communication is facilitated, of course, if the diagnosis is simple and straightforward.

Another unique advantage of a psychiatric rehabilitation diagnosis is that the diagnostic process itself can change clients' perceptions about how much control they have over their own environments. The proponents of skill teaching have argued that success in learning not only improves client skills but also changes clients' expectations and estimates of their own self-efficacy. Similarly, as a consequence of participation in psychiatric rehabilitation diagnosis, the persons with psychiatric disabilities begin to play an active role in their own recoveries.

7

Plans and Interventions

If to do were as easy as to know what were good to do, chapels had been churches and poor men's cottages prince's palaces.

William Shakespeare

Diagnosis is difficult, but it is easier than intervening to address the diagnosis. In times past, mental health practitioners spent too much energy arriving at a psychiatric diagnosis and too little energy planning and intervening with the person based on the diagnosis. A psychiatric rehabilitation diagnosis is only the beginning of the psychiatric rehabilitation approach. If, based on what the practitioner and the client know, little is done, then little has been accomplished.

The psychiatric rehabilitation plan links the rehabilitation diagnosis to the rehabilitation interventions. Persons with psychiatric disabilities often have several skill and resource deficits. Interventions aim to eliminate these deficits. The rehabilitation plan essentially identifies *who* is responsible for doing *what*, by *when*, for *how long*, and *where*. The rehabilitation diagnosis provides the answer to the question, *Why*.

Based on such criteria as urgency, motivation, and ease of achievement, the practitioner and client assign priorities to skill and resource development objectives. The most urgent and easy to achieve objectives, for which client motivation for achieve-

113

ment is high, are the focus of the interventions. The practitioner identifies a specific intervention for each skill or resource development objective in the plan and a specific person responsible for providing each intervention. The client signs the rehabilitation plan to indicate his or her agreement. Table 7–1 presents an example of a rehabilitation plan.

As illustrated by Table 7–1, several people are involved in providing rehabilitative interventions. The plan operationally defines the rehabilitation team approach. The problem in the past with the team approach has been an inability of the team members to understand the tasks for which the other members are specifically responsible, which frequently resulted in confusing the client and conflict between the team members. Psychiatric rehabilitation cannot have a team approach without a game plan.

The team must be a team in more than name only. Each member must have observable goals for his or her interventions. Obviously the complexity of the plan is related to the number and type of client skill and resource needs as well as to the number of rehabilitation environments under consideration. Many clients require complex and detailed plans.

This chapter focuses on the major issues with respect to planning and implementing skill development and support interventions. First, some research relevant to each of these interventions is reviewed. Second, the issues and principles pertinent to these interventions are presented.

Research Review

Research relevant to the major interventions of psychiatric rehabilitation have been reported in a wide variety of professional journals. For example, research specific to the major interventions have come from human resource development (Carkhuff & Berenson, 1976), social skills training (Hersen & Bellack, 1976), social learning theory (Paul & Lentz, 1977), vocational rehabilitation (Anthony, Howell, & Danley, 1984), and community support (Test, 1984). By drawing together the research from so many fields, support for these two major interventions becomes apparent.

TABLE 7-1 Example: Rehabilitation Plan

Overall Rehabilitation Goal: Mike intends to live in the Northwest apartments by January of next year.

Priority Skill/Resource Development Objectives	Interventions	Person(s) Responsible	Starting Dates Projected	Actual	Completion Dates
Mike says what he thinks about a subject 4 times/week when conversing with residents during social interactions.	Direct skills teaching	*Provider:* House counselor *Monitor:* Mike & group home staff	April 14	April 20	May 29
Mike asks other residents for help 75% of the times/week when he is doing household chores.	Skills programming	*Provider:* Mike *Monitor:* Mike & group home staff	June 1	June 15	July 31
Welfare Department provides Mike with food stamps once per month.	Resource coordination	*Provider:* Department of Social Services *Monitor:* Mike & house counselor	July 1	July 1	July 31

I participated in developing this plan and the plan reflects my objectives. Client's signature: _____

Four types of research are most relevant to these psychiatric rehabilitation interventions:

1. Research analyzing the ability of persons with psychiatric disabilities to learn skills.
2. Research examining the relationship between the skills of persons with psychiatric disabilities and their rehabilitation outcome.
3. Research investigating the relationship between skill development interventions and rehabilitation outcomes.
4. Research investigating the relationship between support interventions and rehabilitation outcome.

Persons with Psychiatric Disabilities Can Learn Skills

At first glance, research investigations as to whether persons with psychiatric disabilities can learn skills may seem superfluous. Common sense tells us they can learn skills. Yet, until the late 1960s and early 1970s, skill development interventions were far from the norm. During that period, a number of studies on skill training were published in a variety of professional journals.

In 1974, Anthony and Margules reviewed these articles and concluded that persons with psychiatric disabilities can in fact learn useful skills (Anthony & Margules, 1974). Specifically, the studies reviewed at that time showed that persons with psychiatric disabilities could learn a variety of physical, emotional-interpersonal, and intellectual skills. For example, in the physical area of functioning, skill training programs have impacted skills in a variety of areas, including personal hygiene (Harrand, 1967; Retchless, 1967; Scoles & Fine, 1971; Weinman, Sanders, Kleiner, & Wilson, 1970), cooking (Scoles & Fine, 1971; Weinman et al., 1970), use of public transportation (Harrand, 1967), use of recreational facilities (Harrand, 1967), use of particular job tools (Shean, 1973), and physical fitness (Dodson & Mullens, 1969). In the emotional-interpersonal area of functioning, skill training programs have increased interpersonal skills (Ivey, 1973; Pierce & Drasgow, 1969; Vitalo, 1971) socialization skills (Bell, 1970; Weinman et al., 1970), self-control skills (Cheek & Mendelson, 1973; Rutner & Bugle, 1969), selective reward skills (Swanson

& Woolson, 1972), and job interviewing skills (McClure, 1972; Prazak, 1969). Last, in the area of intellectual functioning, skill training programs have increased money management skills, (Weinman et al., 1970), job-seeking skills (McClure, 1972), and job-applying skills (McClure, 1972; Safieri, 1970).

Many of these early studies trained persons with psychiatric disabilities who were long-term inpatients with a lengthy history of symptomatic behavior. It is now widely accepted that persons with psychiatric disabilities can learn skills, and that neither chronicity nor symptomatology prevents skill learning. The results of these early studies are thus seen as supportive of a psychiatric rehabilitation approach, with its emphasis on assessing and improving persons' skills.

The concept of skills in the psychiatric rehabilitation approach is much broader than daily living skills. For example, the skill of symptom management can be taught. Breier & Strauss (1983) conceptualize this skill as a three-step process: "the detection of an unwanted behavior or its antecedent, the evaluation of such behavior as a warning or abnormal, and the employment of a control strategy" (p. 1144). Most problematic to assess and teach are intrapersonal and interpersonal skills. Yet these are the types of skills with which persons with psychiatric disabilities most often need help.

In a review of social problem solving skills in schizophrenia, Bellack, Morrison, and Mueser (1989) concluded that persons with schizophrenia are critically deficient in their ability to communicate and make themselves understood. They report that "the evidence for this impairment is much stronger than the evidence for a deficit in problem solving skill. It would appear that the communication impairment is more central to their interpersonal difficulties than deficits in problem solving ability and the inability to make their desires and reasoning understood would seriously compromise the ability of patients with schizophrenia to use the strategies taught in problem solving training" (pp. 111-112). The Center for Psychiatric Rehabilitation (1989) recently studied the efficacy of direct skills teaching, a skill training technology, on the acquisition, application, and use of selected interpersonal skills. Using a single-subject experimental design, replicated across subjects, a number of unique interpersonal skills were assessed

117

and taught. These skills included (1) stating personal thoughts; (2) refusing requests; and (3) sharing negative feelings. Analyses indicate that the technology of direct skill teaching positively and significantly affected the use of these interpersonal skills.

Skills Relate to Rehabilitation Outcome

Other research studies central to the practice of psychiatric rehabilitation show a relationship between the skills of the client and measures of rehabilitation outcome. Typically, these studies have investigated the relationship between work adjustment and interpersonal skills and vocational outcome measures. However, one study (Dellario, Goldfield, Farkas, & Cohen, 1984) did report a significant relationship between ratings of life skills and discharge from a psychiatric inpatient setting, that is, those patients rated as most skilled were most apt to be discharged.

The studies investigating the relationship between client skills and vocational outcome are remarkably consistent. Every study that assessed work adjustment skills found them to be significantly related to future work performance (Bond & Friedmeyer, 1987; Cheadle, Cushing, Drew, & Morgan, 1967; Cheadle & Morgan, 1972; Distefano & Pryer, 1970; Ethridge, 1968; Fortune & Eldredge, 1982; Green et al., 1968; Griffiths, 1973; Miskimins, Wilson, Berry, Oetting, & Cole, 1969; Watts, 1978; Wilson et al., 1969). When an overall measure of work adjustment skills was calculated, the total score was always predictive of future vocational performance (Bond & Friedmeyer, 1987; Cheadle et al., 1967; Cheadle & Morgan, 1972; Distefano & Pryer, 1970; Ethridge, 1968; Griffiths, 1973). The ratings of work adjustment skills in each of the studies were done by vocational counselors, occupational therapists, or work supervisors. The settings for these ratings were various sheltered or simulated work environments.

Ratings of interpersonal or social skills have also been found to predict vocational performance (Green et al., 1968; Griffiths, 1974; Gurel & Lorei, 1972; Miskimins et al., 1969; Strauss & Carpenter, 1974; Sturm & Lipton, 1967). Again the data consistently suggests that knowledge of social functioning can be used to make inferences about clients' future vocational performance.

The concept of social functioning is described differently

in each study. For example, Green and associates (1968) rated the ability to initiate social contacts with other patients and staff. Miskimins and associates (1969) rated social skills, and Gurel and Lorei (1972) reported a significant relationship with the raters' estimate of restricted psychosocial functioning. Griffiths (1973) evaluated such items as getting along well with other people and communicating spontaneously, whereas estimates in the Strauss and Carpenter (1974) study were of personal-social relations, that is, meeting with friends or participating in activities with social groups. Thus, in spite of the wide range of items used to measure social functioning, the studies were remarkably similar in finding a relationship between social ability and future vocational performance.

Because client skills, and not client symptoms, are the focus of the psychiatric rehabilitation approach, it is interesting to contrast the preceding body of research with studies that have examined the relationship between client symptoms and rehabilitation outcome. As reported in chapter 2, many studies have illustrated the low correlations between a variety of assessments of psychiatric symptomatology and future work performance.

Skill Development Interventions
Impact Rehabilitation Outcome

Central to the psychiatric rehabilitation approach is the assumption that skill development interventions will increase the capacity to live, learn, socialize, and work more independently and effectively. In their early review of a number of databased studies, Anthony and Margules (1974, p. 104) indicated that the research suggested that "persons with psychiatric disabilities can learn skills; and that these skills, when properly integrated into a comprehensive rehabilitation program which provides reinforcement and support for the use of these skills in the community, can have an effect on the community functioning."

This literature has recently been reviewed again (Anthony, Cohen, & Cohen, 1984; Liberman et al., 1986). Although the majority of studies of skill development interventions are part of the behavioral psychology literature and are not focused on persons with severe psychiatric disabilities (Hersen, 1979), several studies are relevant, with new studies appearing regularly (Liberman, Mueser, & Wallace, 1986; Wong et al., 1988). Anthony, Cohen, and Cohen (1984) identified studies undertaken with persons with severe psychiatric disabilities. Besides demonstrating that clients can in fact learn skills, many of these studies examined additional types of outcomes. For example, Vitalo (1979) compared a skill training approach with a medication group for long-term outpatients and reported significant increases in the skill training group in terms of the number of new friends and number of new activities in which the clients were involved.

Some vocational outcome studies have examined the effects of training in job-seeking skills (Azrin & Philip, 1979; Eisenberg & Cole, 1986; McClure, 1972; Stude & Pauls, 1977; Ugland, 1977), decision-making skills (Kline & Hoisington, 1981), and occupational and work adjustment skills (Rubin & Roessler, 1978). Each study reported improved employment outcomes for the groups of persons with severe psychiatric disabilities who received skill training. For example, Eisenberg & Cole (1986) reported 61% of their clients trained in job-seeking skills obtained employment versus 12% for a matched control group.

Support Interventions Impact Rehabilitation Outcome

In addition to skill development interventions, the other major psychiatric rehabilitation intervention focuses on increasing the support in the person's environment. In essence, two types of research studies have manipulated the environment. The first type of study used both client skill development and support interventions in concert; thus, it is impossible to identify the unique contributions of either intervention to client outcome. The second type of study researched intervention strategies more exclusively focused on increasing the support in the environment in which the client functions. The studies focusing more exclusively on support interventions are examined first.

Support interventions attempt to provide the client with supportive persons, supportive places, supportive activities, and/ or supportive things. A supportive person offers support through performing required behaviors or different roles (i.e., advocate, case manager, counselor, and/or advisor). Improving the support provided by people, places, activities, or things focuses on accessing or modifying resources within the environment. (i.e., sheltered or supported work and living settings, special transportation, spending money, discharge programs). The purpose of distinguishing among supportive persons, supportive places, supportive activities, and supportive things is simply to highlight the different ways environmental modifications occur. In practice, these modifications often occur simultaneously.

The main identifying feature of support interventions, as distinguished from skill development interventions, is that they do not attempt to change the client's behavior. Rather, the attempt is to support and accommodate the client's present level of functioning. The early studies by Katkin, Ginsburg, Rifkin, and Scott (1971) and Katkin, Zimmerman, Rosenthal, and Ginsburg (1975) and later studies by Cannady (1982) and Schoenfeld, Halvey, Hemley van der Velden, and Ruhf (1986) have clearly demonstrated the positive impact on client outcome of supportive persons. For example, Cannady (1982) employed citizens from the discharged patients' rural neighborhood to function as supportive case workers. Results indicate that over a 12-month period inpatient days were decreased by as much as 92%.

Witheridge, Dincin, and Appleby (1982) have reported on the use of a support team for persons with psychiatric disabilities at high risk for readmission. Working out of the clients' homes and neighborhoods, this team intended to develop an individualized support system for each client. Of the original 50 participants, 41 remained in the program. One-year follow-up data indicated that days hospitalized decreased from 87.1 to 36.6 per individual.

A study by Stickney, Hall, and Gardner (1980) investigated the effects of introducing both a supportive person *and* a more supportive environment, separately and in combination, at the time of the person's hospital discharge. They studied the impact of 4 pre-discharge strategies that differed in the level of person and environmental support. The goals of the discharge plans were to increase the use of the community mental health center and decrease recidivism for 400 patients discharged from a state psychiatric hospital. The results of the study demonstrated the impact of increasing person and environmental support on both client compliance with referral and 1-year hospital recidivism rates. With minimal environmental support, the referral compliance and recidivism percentages were 22% and 68%; with increased person support, 36% and 39%; with increased person and environmental support, 75% and 28%. Thus, whenever an added element of support was introduced the percentage of referral compliance increased and the recidivism rate decreased.

Other studies (Valle, 1981; Weinman & Kleiner, 1978) have investigated the relationship between support persons and rehabilitation outcome. However, in each of these studies the effect of the support person could not be differentiated from the impact of skill training. Valle (1981) investigated the relationship between levels of interpersonal skills of the supportive counselor and rehabilitation outcome. He reported that the relapse rate of alcoholic clients was significantly related to the level of interpersonal skills of their counselors. In other words, the best predictor of client drinking behavior at the 6-, 12-, 18-, and 24-month follow-ups was the interpersonal skill level of the counselor.

Weinman and Kleiner (1978) did not measure the interpersonal skills of the support persons. Their study compared the effectiveness of community-based enablers with two hospital-based conditions: socioenvironmental therapy and traditional hospital treatment. The enablers' major roles were to teach their patients skills and escort them to various community resources. Results of the project concluded that this combination of person support plus skill training was superior to one or the other of the hospital-based treatment approaches in terms of recidivism, self-esteem, and instrumental role performance.

Some case management studies have combined skill development and support interventions. (See chapter 10 for an overview of case management.) Ballantyne (1983), Wasylenki and associates (1985), and Goering, Wasylenki, and associates (1988) have reported results from a case management study based directly on the philosophy and technology of psychiatric rehabilitation presented in this text. Community rehabilitation workers were trained to provide interpersonal support, skill training, and case management to patients being discharged from the hospital. These patients were compared to a matched group of patients receiving traditional discharge planning and aftercare. Compared to the control group, the experimental group were discharged an average of 21 days sooner and participated in approximately twice as many aftercare services. Client perception of the program indicated high levels of involvement and satisfaction. At 6-month follow-up, more than 75% of the clients were still involved in the program; 87% reported feeling understood, and 88% reported feeling like a partner in the rehabilitation planning process (Ballantyne, 1983). Client outcome data at 6 and 24 months showed improved client functioning over time. By 24 months, the experimental group was significantly better than the control group on measures of instrumental role functioning, occupational and housing status, and social isolation (Wasylenki et al., 1985; Goering, Wasylenki, et al., 1988).

Last, several inpatient studies have combined skill training interventions with a variety of adjunctive services, including community support once the patient is discharged. Undoubtedly the most well known study is Gordon Paul's research (1984), which compared a traditional hospital unit with a milieu therapy approach and a social learning approach. Among the many process and outcome measures used to evaluate these programs were community tenure figures that suggested the social learning approach was superior to milieu therapy, which in turn was superior to the traditional inpatient program. Other much less controlled and comprehensive evaluations have been performed on inpatient skills-oriented programs; each has reported favorable outcome on measures of community tenure and community functioning (Becker & Bayer, 1975; Heap, et al., 1970; Jacobs & Trick, 1974; Waldeck, Emerson, & Edelstein, 1979).

In summary, a variety of empirical studies conducted by researchers from different mental health and rehabilitation disciplines suggest that:

1. Persons with severe psychiatric disabilities can learn skills.
2. The skills of persons with psychiatric disabilities are positively related to measures of rehabilitation outcome.
3. Skill development interventions improve psychiatric rehabilitation outcome.
4. Support interventions improve psychiatric rehabilitation outcome.

Interventions Issues and Principles

As the research indicates, the acquisition of skills is not a major problem in rehabilitating persons with psychiatric disabilities. They can and do learn skills when effectively taught. However, the acquisition, or learning, of skills is more easily demonstrated than the application, or performance, of skills in the relevant environment. *Learning* relates to the question, Can the client acquire the skill? while *performance* asks the question, Will the client use the skill? (Goldstein, 1981).

Skill performance, application, or generalization is not a unique issue for skill development interventions. It has been estimated that the average generalization and maintenance rates for all psychotherapies combined is 14% (Goldstein & Kanfer, 1979). Nevertheless, this major issue must be directly confronted by skill development interventions.

Based on a review of the skill-use literature, Cohen, Ridley, & Cohen (1985) have summarized the 11 principles that should be incorporated into a skill development intervention to maximize the possibility of generalization:

1. Use the natural reinforcers present in the relevant environment to reward appropriate responses in the training environment.
2. Provide support services to the client in the relevant environment.
3. Teach support persons to use the skill of awarding selective rewards in the relevant environment.
4. Teach the client to identify intrinsic motivation as a replacement for extrinsic reward.
5. Increase the delay of reward gradually.
6. Teach skill performance in a variety of situations.
7. Teach variations of skill use in the same situations.
8. Teach self-evaluation and self-reward.
9. Teach the rules or principles that underlie the skill.
10. Use gradually more difficult homework assignments.
11. Involve the client in setting goals and selecting intervention strategies.

One skill teaching technology that incorporates these principles is direct skills teaching (DST). DST is a systematic method of outlining the knowledge needed to learn any skill, developing a structured lesson plan to teach each component behavior, and involving the client in practicing each behavior individually and all the behaviors together (Cohen, Danley, & Nemec, 1985).

Providing ongoing support for skill use is a major method for improving skill generalization. The two types of psychiatric rehabilitation interventions (skill development and support) are inextricably linked. Skill teachers support skill use (Cohen, Danley, & Nemec, 1985). The use of skills by persons with psychiatric disabilities can be improved through simple social reinforcements as opposed to elaborate token economies or reinforcement schedules (Armstrong, Rainwater, & Smith, 1981). Persons with psychiatric disabilities do not respond well to mere token reinforcement. Social reinforcement is less expensive, more acceptable and normal, and can be more effective in producing the desired skill use in the targeted environment.

Similarly, practitioners providing a support intervention can help clients to *learn* to access natural supports rather than simply providing these supports. Along these lines, Mitchell (1982) has reported data suggesting that persons' skills are positively related to the size of the persons' support networks. Individuals can be taught to take an active role in the development of their own support network. Mitchell's (1982) data indicate that the level of independence and the level of peer support are positively related. The more support one feels, the more independently one can act. The principle is that if individuals are taught to get the support they want, they can function more independently.

Other studies have looked at the relationship between support and outcome (Greenblatt, Bererra, & Serafetinides, 1982). Dozier, Harris, and Bergman (1987) found that moderate levels of network density, or the extent to which network members know one another, were associated with fewer days in the hospital. In a study conducted in a Fountain House type clubhouse, Fraser, Fraser, and Delewski (1985) reported that reduced hospitalization was correlated with increases in the number of mental health professionals in the clubhouse members' networks. Sommers (1988) found that significant others' attitudes, that is, high demands for socially appropriate behavior, were correlated with instrumental performance. Baker, Kazarian, Helmes, Ruckman, and Tower (1987) found that readmitted patients rated their second most influential person higher on overprotection and criticism scales and lower on the care scale than did nonrecidivists.

All types of support are critical but *the provision of personal support to persons who are psychiatrically disabled is most critical*. To be supported by others is a natural human desire. This desire is especially critical for persons who are psychiatrically disabled because impairment may have caused them to withdraw from activities that provide a social outlet, that is, work, school, family activities (Harris, Bergman, & Bachrach, 1987). The support networks of persons with psychiatric disabilities are characterized by small size, a lack of reciprocity, domination by kin (Weinberg & Marlowe, 1983), and inflexibility and instability (Morin & Seidman, 1986).

Research into support interventions is plagued by conceptual and methodological shortcomings (Lieberman, 1986; Starker, 1986; Thoits, 1986). The concepts of support, support networks, and support systems do not have a clear definition. Uniform and reliable instruments are lacking. Despite these deficiencies, however, the idea of a support intervention as an effective rehabilitation intervention has taken hold.

One of the more useful definitions of support has been provided by Kahn and Quinn (1976), who describe support as an interpersonal transaction consisting of expression of positive affect, affirmation, and aid. *Positive affect* includes liking, admiration, respect, and other kinds of positive evaluation; *affirmation* includes endorsement of an individual's perceptions, beliefs, values, attitudes, or actions; *aid* includes such things as materials, information, time, and entitlements. Peers, practitioners, and family members provide varying levels and types of personal support. The principle is that if increased independent functioning is a client goal then some person or persons must be providing positive affect, affirmation, and aid.

Based on the research literature, possible interventions to

improve clients' supports have been suggested (Harris & Bergman, 1985; Marlowe & Weinberg, 1983). These interventions include:

1. Modification of the pre-existing network (family and friends) to increase their expressions of support.
2. Development of additional network members by using volunteers.
3. Development of additional network members by paying volunteers.
4. Strengthening involvement in natural networks (e.g., church, clubs) already in the community.
5. Development of a group of people who have similar problems into a support network.
6. Strengthening the links between existing network members.
7. Using larger networks for a crisis intervention function.
8. Expanding the functions of the network.

A fundamental principle underlying any support intervention is that the person with the disability, not the practitioner, must perceive the intervention to be supportive. Support, like beauty, is in the eye of the beholder. Not all personal relationships are supportive and beneficial (George, Blazer, Hughes, & Fowler, 1989). The affect, affirmation, or aid must be considered to be beneficial by the recipient of the support. Just as what stresses one person is different from what stresses another person, so it is with support.

Concluding Comment

It should come as no surprise that even though the psychiatric rehabilitation process of diagnosis, planning, and intervening can be described relatively simply, it is not simple to provide. Expert personnel, effective programs, and well-designed service systems must be in place in order to increase the chances of persons reaching their rehabilitation goals. The remainder of the text focuses on how these people, programs, and systems can facilitate the psychiatric rehabilitation process.

Personnel

8

I expect to pass through life but once. —If therefore, there be any kindness I can show, or any good thing I can do to any fellow-being, let me do it now, and not defer or neglect it, as I shall not pass this way again.

William Penn

The three dimensions of personnel, programs, and systems provide a conceptual framework for understanding what contributes to rehabilitation outcome. In essence, the rehabilitation outcome of persons with severe psychiatric disabilities is influenced by the effectiveness of the personnel, programs, and service systems with which the client interacts. By analyzing personnel, programs, and systems, the rehabilitation process can be more effectively investigated, used, and changed. The three units of analysis are:

1. The skills, knowledge, and attitudes of the various personnel who interact with the client (e.g., counselors, psychiatrists, family members, other clients).
2. The programs used by the personnel (e.g., peer companion programs, supported work programs, independent living skills programs).
3. The service systems that support the people and programs (e.g., vocational rehabilitation agencies, state departments of mental health).

This chapter focuses on the personnel dimension.

Persons who are psychiatrically disabled interact with many people who can have a significant impact on their rehabilitation outcome. Practitioners (credentialed and noncredentialed), family members, and other consumers (clients, ex-patients) usually have the greatest impact. The most significant factor is not the role of the personnel, but rather their ability to perform certain functions such as the tasks performed when diagnosing, planning, and intervening in a helpful way. No matter what the title, credential, or role, the personnel must have the requisite skills, knowledge, and attitudes to perform these activities:

1. *Connect* with clients to develop a close bond.
2. Help clients *set self-determined rehabilitation goals*.
3. Help clients *assess* their personal skills and environmental supports in relation to their overall rehabilitation goals.
4. Help clients *plan* to develop the skills and/or resources they need.
5. Help clients *learn* the new skills they need to learn.
6. Help clients *use* the skills they already possess.
7. Help clients *link* with the resources they need.
8. Help clients *modify* their resources to improve the support provided.
9. Give clients the ongoing *personal* support they need.

Each activity needs to occur in the diagnostic, planning, and intervention process. Each activity ensures that clients obtain the skills and resources they need to reach their goals. The second and third activities are performed in the diagnostic phase. The fourth activity is performed in the planning phase. Activities 5 through 8 are part of the intervention phase, and activities 1 and 9 are part of any and all phases of the psychiatric rehabilitation process.

The personnel varies in how systematically they perform these activities. At times they may not even be aware that they are performing them. Nevertheless, persons with psychiatric disabilities need to interact with personnel capable of performing these activities when needed.

The importance of the helping person's interpersonal skills

cannot be overstated. Interpersonal skills, obviously essential for activities 1 and 9, underlie *all* activities in the rehabilitation process. Such skills as demonstrating understanding, self-disclosing, and inspiring (Cohen et al., 1989) relate to outcomes in the practice of counseling (Carkhuff & Anthony, 1979), teaching (Aspy, 1973), and drug rehabilitation (Valle, 1981). Interpersonal skills can be practiced by professionals, family members, and consumers.

Goering and Stylianos (1988) argue that the efficacy of current rehabilitation interventions is due in part to the relationship that develops between the client and the practitioner. One of the most potent ingredients of effective rehabilitation is the strength of the therapeutic alliance (Goering & Stylianos, 1988). Such a conclusion is consistent with the emphasis on interpersonal relationships in most of the activities of psychiatric rehabilitation. The diagnosis/planning/intervention process of rehabilitation equally emphasizes interpersonal relationships and the activities of goal setting, assessment, planning, and skill and resource development.

Credentialed Professionals

Credentialed professionals (i.e., persons with a PhD, MSW, OTR, CRC, MS, MA, RN, or MD) who work in this field must perform many activities for which traditional professional training programs have *not* prepared them (Anthony, 1979) and for which they may have little interest (Stern & Minkoff, 1979). The prevailing attitude among many mental health professionals is that working with persons with long-term psychiatric disabilities is unrewarding, nonprestigious, and hopeless (Minkoff, 1987).

The lack of training in the technology needed to work effectively with persons with severe psychiatric disabilities (Minkoff, 1987) is particularly apparent when credentialed professionals are asked to perform such skills as those used to conduct a functional or resource assessment, develop a rehabilitation plan, teach skills, construct a skills-use program, coordinate resources, and/or modify resources (Marshall, 1989). Psychiatric rehabilitation practitioners must master this technology in order to guide clients through the psychiatric rehabilitation process (Anthony, 1979; Spaulding, Harig, & Schwab, 1987). Chapter 5 discusses this technology and lists the activities and skills that someone trained in this technology can perform.

A focus on technology serves to mitigate the importance of degrees and titles. Rather than distinguish between case manager, therapist, counselor, MD, MSW, RN, and so on, it is more sensible to make distinctions between professionals in terms of what they do well and ensure that the rehabilitation team is comprised of persons who as a group can perform all the necessary skills to guide someone through the rehabilitation process. It makes no sense to build a team by credentials or titles, so that programs have, for example, 2 social workers, 3 mental health workers, 1 psychologist. Except when a certain type of care is legally mandated (e.g., medical and nursing care, psychological testing), personnel should be organized into a functional team so that all the requisite activities are performed.

In a review of professional training programs, Anthony, Cohen, and Farkas (1988) concluded that the four core disciplines of psychology, social work, psychiatric nursing, and psychiatry have been, at best, reluctant to train their professionals to provide services to persons with long-term psychiatric disabilities (Friday, 1987). Bevilacqua (1984) has criticized the social work, psychology, and nursing disciplines for adopting an office-based, private practice model of medicine that removes these disciplines from the environmental and social problems of persons with psychiatric disabilities. As a result, little attention is paid to long-term mental illness in the curricula of the professional schools. Rapp (1985) and Gerhart (1985) have pointed out three obstacles to training common to all the professional disciplines:

1. Lack of interested and experienced faculty.
2. Lack of suitable field agencies and supervisors.
3. Lack of student interest in working with the population.

In 1984, Talbott examined the previous 5 years of educational developments in training professionals to work with the long-term mentally ill (Talbott, 1984). He registered surprise that although policy and service delivery issues pertaining to persons with long-term mental illness sparked much interest, there was no similar interest in educational and training issues. Talbott believes psychiatric residency programs pay too little attention to content areas such as functional assessment, the elements of a community support system, and case monitoring over a long period of time. With respect to continuing medical education programs, he recommends more exposure to, among other things, successful psychiatric rehabilitation programs in the community. Cutler's model of residency training (Cutler, Bloom, & Shore, 1981) and several others are notable exceptions, but most training programs have not adopted the necessary content or field experiences. Stein, Factor, and Diamond (1987), in agreement with Talbott, identified only a handful of residency training programs that provide sufficient course work and experience in treating persons with long-term psychiatric disabilities. Young psychiatrists do not find persons with long-term psychiatric disabilities an attractive population to treat.

Social work students find the psychodynamic model more attractive to study than the more recently developed psychiatric rehabilitation approach. The students, as well as their professors, are more comfortable with the private practice model of social work. Rubin (1985) has reported survey data indicating that persons who are psychiatrically disabled are rated by students as the least appealing group with whom to work.

Davis (1985) has focused a good deal of his analytical work on the university preparation of mental health professionals for work with persons with long-term psychiatric disabilities. He examined the characteristics of clients in the Virginia mental health system and the implication of this data for professional training. His data indicated a lack of relevant curricula and practica, particularly in psychology and social work (Davis, 1985). With respect to rehabilitation counselor training, Weinberger and Greenwald (1982) surveyed 59 accredited graduate programs in rehabilitation counseling and reported that only 7 offered a specialty in psychiatric rehabilitation. McCue and Katz-Garris (1985) surveyed 580 practicing rehabilitation counselors about their training needs. The highest ranked training needs were (1) translating psychiatric diagnosis into functional, behavioral terminology; (2) career counseling and realistic goal planning; and (3) teaching job-seeking skills and meeting job placement needs.

Despite the absence of an adequate preservice curriculum, the knowledge and research base for developing such a curriculum has expanded exponentially during the last several decades (Anthony & Liberman, 1986). Thus, it is now possible to build a preservice curriculum based on an empirical and experiential literature. A specialty area, for which any of the 4 core disciplines could provide training, can now be defined. This curriculum can be derived from what is currently known about how best to help persons with long-term psychiatric disabilities function more effectively in the communities of their choice (Anthony, Cohen, & Farkas, 1988).

In essence, such a curriculum can evolve from a knowledge of the variables that relate to positive client outcomes. This empirically based knowledge about the correlates and causes of client outcome implies certain activities a practitioner should perform to improve client outcome; the activities in turn imply certain skills the practitioner needs to perform the activities. With improved client outcome as the main reason for a particular curriculum, a preservice curriculum can help the practitioner develop skills relevant to performing those tasks that appear related to positive client outcomes.

Curriculum Intensity

Preservice training programs will vary in terms of the intensity of the professional training experience. Some programs directly teach psychiatric rehabilitation skills. These programs explain the steps for skill performance, demonstrate skill performance, and offer supervised opportunities for practicing skill performance. Other programs simply provide general knowledge about psychiatric rehabilitation but provide no skills training. Still other programs provide both the knowledge and supervised fieldwork but neglect systematic instruction in the skill.

Cohen (1985) has proposed a way of categorizing preservice programs that teach students about the rehabilitation of persons with long-term psychiatric disabilities. Curricula of these programs can be categorized with respect to their level of intensity (see Table 8–1).

A preservice program assessed at the exposure level of intensity increases student awareness of psychiatric rehabilitation knowledge by means of didactic courses or parts of courses. A

134

TABLE 8–1 Level of Training Intensity in Preservice Training Programs

	Exposure	*Experience*	*Expertise*
Mission:	Develop knowledgeable practitioners	Develop experienced practitioners	Develop skilled practitioners
Goal:	Increased knowledge	Improved attitudes; increased knowledge	Improved skills and attitudes; increased knowledge
Method:	Didactic coursework	Supervised fieldwork; didactic coursework	Skill-based courses, supervised fieldwork, didactic coursework
Evaluation:	Written tests and papers	Supervisor's evaluation; number of client contact hours, types of client contact; written tests and papers	Taped ratings of skill performance; client satisfaction; supervisor's evaluations; number of client contact hours; types of client contact; written tests and papers

Adapted from: Anthony, Cohen, Farkas (1988)

curriculum at the experience level of intensity supplements the didactic material with supervised fieldwork or internships. The expertise level is reserved for curricula with skills building courses in addition to the didactic coursework and fieldwork experience (Anthony, Cohen, & Farkas, 1988).

Clearly, the field needs more expert practitioners. It is one thing to provide exposure to psychiatric rehabilitation as a specialty area. It is another thing to add some experience. It is quite another to develop expert practitioners. Expert practitioners can demonstrate their skills by means of audio or videotapes of their sessions with clients (Rogers et al., 1986). Their skill development can be observed and their new learning measured from pre- to posttraining (Anthony, Cohen, & Farkas, 1987). Expert practitioners, unlike practitioners who possess only knowledge and experience, can document their ability to perform certain skills. A curriculum that trains expert practitioners adds credibility to the field of helping persons with psychiatric disabilities (Minkoff, 1987).

An example of an expertise-level curriculum is Boston University's MS degree program in psychiatric rehabilitation (Farkas, O'Brien, & Nemec, 1988). Students in this program take nine courses on rehabilitation knowledge in which their learning is assessed by written tests and papers (exposure level). They participate in clinical settings with several persons with long-term psychiatric disabilities and are rated and evaluated by supervisors (experience level). Students take six courses in which skill performance is taught and evaluated. The skills building courses use multimedia training technology packages (M. R. Cohen et al., 1985, 1986, 1989) including trainer guides, student reference handbooks, and audiovisual demonstrations of skill use. Students' actual performances in their fieldwork with persons with psychiatric disabilities are observed and rated by means of audiotaping sessions (expertise level). Skill performance is rated using scales with previously demonstrated reliability and validity (Rogers et al., 1986). Ongoing evaluation of the Boston University program indicates that up to 5 years after graduation, approximately 90% of the graduates work with the long-term psychiatrically disabled population. Student satisfaction questionnaires indicate that 88% of the graduates are very satisfied with the program.

Curriculum Changes

In contrast to the previous lack of curriculum, it appears that within the last several years the academic environment has increasingly offered preservice training in psychiatric rehabilitation. There are several reasons for this current interest in preservice curriculum relevant to persons with long-term psychiatric disabilities. First, changes in services are beginning to influence educators in this direction. Jerrell and Larsen (1985) reported that the services receiving the greatest increases in funding are for persons with severe psychiatric disabilities. State mental health authorities have used unrestricted federal block grants to fund increased services to persons with long-term psychiatric disabilities. Thus, changes in funding priorities influence those who educate practitioners. Likewise, clients and their families are influencing both educators and practitioners to learn more about psychiatric rehabilitation (Lecklitner & Greenberg, 1983; Zipple & Spaniol, 1987). In addition, mental health authorities are influencing universities in their states to make coursework more relevant to this population. As Bevilacqua (1984) has pointed out, the state mental health authority can exert leverage on academia to make their content more pertinent to the needs of this particular client population. Bevilacqua (1984, p. 4) puts it this way:

> With federal funds for professionals' education in mental health diminishing, the states are emerging as important resource replacements. If state government begins to ask the hard questions of how universities are addressing issues dealing with such problems as chronic mental illness, and if state legislatures can be persuaded to provide incentives for increased attention to those issues, an important first step will have been taken.

The efforts of the Council of Social Work Education have been particularly notable. Funded for 3 years in 1982 by NIMH, the Curriculum and Resource Development Project on Chronic Mental Illness worked to strengthen and expand social work education curricula and teaching resource materials on chronic mental illness. Through surveys, conferences, publications, training, and technical assistance, the project has developed and disseminated relevant educational material (Bowker, 1985).

A survey by the Western Interstate Commission for Higher Education (WICHE) examined academia's response to state mental health system needs (Moore, Davis, & Mellon, 1985). This survey collected information on coursework relevant to persons with long-term psychiatric disabilities. They concluded that the interest in strengthening academic programs' ability to prepare graduates to serve client populations that have been assigned high priority by mental health authorities (including persons with long-term psychiatric disabilities) seems strong enough to warrant curriculum development. A survey by the Southern Regional Education Board identified inservice and preservice training programs in psychiatric rehabilitation (Friday, 1987). The report profiled 27 academic institutions that offer some coursework relevant to psychiatric rehabilitation.

In 1986, the Center for Psychiatric Rehabilitation received grant funds to develop and field test a strategy designed to disseminate the psychiatric rehabilitation curriculum developed at Boston University to other preservice programs. Two psychiatry residency programs, 1 joint social work and psychology program, 1 nursing program, 1 social work program, and 1 multidisciplinary program served as field sites for the successful testing of the strategy. Overall, more than 60 preservice graduate training programs requested information about serving as field sites.

In summary, the literature attests to the historical omission of content relevant to persons with psychiatric disabilities in the curricula of most professional training programs. However, current activities suggest that increasing interest in this content is stimulated by the educators themselves, by changes in services, by the desires of consumers and their families, and by encouragement by mental health authorities. As preservice education on psychiatric rehabilitation grows, the hope is it will focus more on training expert practitioners with demonstrated skill in this special area of practice rather than on training merely knowledgeable and experienced practitioners.

Functional Professionals

In addition to those disciplines commonly referred to as professional are the ranks of the functional professional (Carkhuff,

1971). These individuals are often labeled nonprofessionals, para-professionals, companions, volunteers, lay professionals, and sub-professionals (Brook, Fantopoulos, Johnston, & Goering, 1989). The functional professional may be defined as a person who, lacking formal credentials, performs those functions usually reserved for credentialed professionals.

Research indicates that functional professionals can be effective in the mental health field. Carkhuff (1968) reviewed more than 80 articles concerned with the use and effectiveness of functional professionals and concluded that functional professionals can effect significant constructive changes in their clients, can be trained to function at facilitative levels over relatively short periods of time and can be seen, in directly comparable studies, to have clients who demonstrate change as much or more than the clients of credentialed professionals. Another review a decade later confirmed these same conclusions (Anthony & Carkhuff, 1978). Studies and anecdotal reports published in the 1970s (e.g., Katkin et al., 1971, 1975; Verinis, 1970; Weinman et al., 1970; Gelineau & Evans, 1970; Goldberg, Evans, & Cole, 1973; Koumans, 1969; Lewington, 1975) support the earlier conclusions of Carkhuff (1968).

To make the best use of functional professionals, the following points need to be considered (Anthony, 1979):

1. Unselected and untrained lay personnel have *not* consistently demonstrated their effectiveness.
2. Functional professionals (or aides) can have a beneficial effect on psychiatric clients only when they are given additional training and responsibility.
3. The rehabilitation activities in which functional professionals are most adept would appear to be the ability to connect with clients, provide personal support, and teach clients the skills necessary to function more productively in community.

Families

Families are a major resource impacting client rehabilitation outcome. An estimated 50–60% of all patients discharged from psychiatric hospitals return home (Lamb & Oliphant, 1979; Min-koff, 1979), and 50–90% remain in contact with their families (Fadden, Bebbington, & Kuipers, 1987; Lefley, 1987). Due to the importance of families and home environments in the rehabilitation of persons who are psychiatrically disabled, means of maximizing their positive influences need attention. Proponents of community mental health have called for the expansion of services to the family as an essential priority (Davis et al., 1974; Weiner, Becker, & Friedman, 1967). Specifically, families need to acquire the knowledge and skills to cope with their family members who are psychiatrically disabled as they go through the stages of adjustment with them (Vaughn & Leff, 1981). Moreover, in order to ensure that families receive adequate knowledge and assistance, service providers need training in techniques to assist the family in coping effectively.

The problems that families encounter can be addressed from the psychiatric rehabilitation perspective that views family members as varying in the repertoire of the necessary skills and knowledge for helping their family members and coping with their impact on family life. This implies that family members may also develop needed competencies and knowledge through a learning process (Anderson, Hogarty, & Reiss, 1980; Miller, 1981).

Approaches to families consistent with psychiatric rehabilitation provide the family with information and increase their capacity to deal with problematic situations. This type of family intervention is consistent with the approach to families used in physical rehabilitation. Psychiatric rehabilitation attempts to translate human problems into observable, measurable events that can be changed. For example, when a family member is unable to deal with a stressful situation in relation to a disabled relative, the rehabilitation approach, instead of concentrating on the childhood patterns that contributed to their present attitudes toward life, or merely prescribing medication to either the relative or family member, focuses on how the family member can acquire knowledge or skills to deal more effectively with the stressful situation.

Families have indicated a need for specific services to help them deal with their disabled family members. First and foremost is the recurring theme of insufficient knowledge and information regarding mental illness, its treatment, availability of services, and practical management techniques (Creer & Wing, 1974; Doll, 1976; Evans, Bullard, & Solomon, 1961; Hatfield, 1979; Hibler, 1978). Especially emphasized are the needs for assistance with management of behavior, tasks of daily living (Creer & Wing, 1974; Hatfield, 1978), and emotional support (Creer & Wing, 1974; Doll, 1976; Hatfield, 1978; Lamb & Oliphant, 1979).

Assistance and support for family members are related to the necessity for teaching service providers how to be most helpful to the family (Starr, 1982). Families have strongly pointed out that service providers are not always helpful and are oftentimes unsupportive when family members want to be a part of the process (Appleton, 1974; Creer & Wing, 1974; Doll, 1976; Hatfield, 1978). Providers must offer families supportive services to enable them to function optimally when a family member returns. Such services may involve providing practical knowledge of mental illness, teaching practical behavior management strategies, and providing information about resources.

According to Spaniol, Zipple, and Fitzgerald (1984) families do not view psychotherapy as giving them what they want. What families appear to want from therapy is different from what professionals want to provide in therapy (Hatfield, Fierstein, & Johnson, 1982). Families want practical advice and information whereas professionals like to focus on family dynamics and emotional expression. Families also feel that the attitudes and assumptions of therapists tend to blame families for the disability and its consequences. Because of high dissatisfaction with professionals, families consistently tend to rate therapy as a low priority across research studies (Hatfield, 1983; Spaniol et al., 1984). Yet, even though families are displeased with the kind of services provided, they still tend to look to professionals for support and practical assistance. As a result, the level of dissatisfaction with the amount of contact with professionals is also high. Sixty percent of families surveyed said they wanted more contact (Spaniol et al., 1984). The average number of contacts reported was less than one per month. Because the amount of contact is quite limited and the quality of the available contacts is not high, families feel an acute sense of abandonment from professionals when their family member returns to the community. Grella and Grusky (1989) found that family satisfaction with services was related to the amount of interaction families had with the case manager, and this variable was more important to family satisfaction than were service system quality and coverage.

New approaches to working with families, loosely categorized as family management or psychoeducational approaches have been increasing in popularity during the 1980s (e.g., Anderson, Hogarty, & Reiss, 1981; Byalin, Jed, & Lehman, 1982; Falloon et al., 1982; Goldstein & Kopeiken, 1981; Leff et al., 1982; Schooler et al., in press). Some of these new approaches have based their interventions on the idea that families who are high in expressed emotion (EE) particularly need these interventions because research has shown high EE measures to be related to the relapse of the family member who is psychiatrically disabled (Brown, Birley, & Wing, 1972; Vaughn & Leff, 1976; Falloon et al., 1982). The level of expressed emotion is inferred from a score on the Camberwell Family Interview and essentially consists of an assessment of the number of criticisms made by family members, the presence of hostility, or ratings of the family members' emotional overinvolvement. Yet there are real questions as to whether the EE concept is useful or necessary; certainly family intervention can proceed successfully without the concept (Hatfield, Spaniol, & Zipple, 1987). Researchers still do not know what EE actually measures, why persons from families who are assessed as being high EE relapse more often (Mintz, Liberman, Miklowitz, & Minty, 1987), or whether the link between EE and relapse is a causal one (Hogarty et al., 1988; McCreadie & Phillips, 1988; Parker, Johnston, & Hayward, 1988).

Zipple and Spaniol (1987) have suggested that four types of family interventions meet all or some of the most critical needs of families. They categorize these existing interventions as:

1. Educational interventions designed primarily to provide information.
2. Skill training interventions designed primarily to develop skills.
3. Supportive interventions designed primarily to enhance the family's emotional capacity to cope with stress.

4. Comprehensive interventions that incorporate information, skill training, and support in a single intervention.

It should be noted that most models use elements of more than one of these approaches. However, these conceptual divisions are useful in that they provide a clear image of the central goal of the intervention.

In their review of the outcome studies that have evaluated these various family interventions, Zipple and Spaniol (1987) concluded that each type of intervention seems to be equally effective in reducing the relapse rate of the mentally ill family member. They suggest two possible explanations for the similarity in the effectiveness of these interventions. First, global outcome indicators such as relapse or recidivism rates are too broad to capture significant differences between the interventions. Using outcome measures of family dissatisfaction, client satisfaction, client level of functioning, family level of functioning, and stress level may demonstrate some differences between the interventions in terms of outcome. Second, the interventions share a core group of common features resulting in similar outcomes. Each intervention attempts to engage the family in a partnership and gives the family some control over the intervention. Each intervention attempts *not* to blame the family for the disability of the member with a psychiatric disability. Each intervention provides either skills, information, or support, or some combination of these services designed to help families cope more effectively with their relative. The comparable outcomes of all these interventions can be explained by arguing that each one involves a range of strategies that may actually be independent of a specific causal model of family distress or patient relapse, but that somehow relates directly to what families have been saying they want or need from professionals. Professionals would be well advised to learn and use these interventions in addition to or in place of more traditional therapeutic interventions.

However, the most dramatic innovation in the role of the family has not been the development of a new family intervention or new theory, but rather the development of the National Alliance for the Mentally Ill (NAMI), a national family advocacy and self-help group. NAMI was not developed by professionals in response to the needs of families. Instead, NAMI was developed by individuals with psychiatrically disabled relatives as a self-help and advocacy group (Hatfield, 1981). Since its official founding in 1979 NAMI has grown dramatically to include hundreds of local chapters representing more than 80,000 members.

NAMI is apparently meeting many needs of families more successfully than many professionals do. It provides families with a powerful vehicle for advocating changes in the mental health system. Family members also discovered much comfort in mutual support (Hatfield, 1981). Joining together to share experiences of common problems seems to be useful for many individuals and is the basis for many self-help groups (Killilea, 1976, 1982).

NAMI is a large, successful group, growing in size and influence. Hatfield (1979) surveyed 79 NAMI and NAMI-like support groups and found that a large majority of the members were satisfied with their groups. Many chapters are quite actively involved in political advocacy (Straw & Young, 1982), and some chapters are beginning to provide direct services to persons with psychiatric disabilities (Shifren-Levine & Spaniol, 1985). Support/advocacy functions for members remain the core of most groups, however.

Consumers

Clients can obviously influence one another, which can be positive in terms of psychiatric rehabilitation outcome. The impact on psychiatric rehabilitation outcome of ex-patient groups, drop-in centers, or clubs is gradually being recognized. Ex-patients have indicated that these types of groups are helpful (Chamberlin, 1978; Peterson, 1979; Zinman, 1982). Practitioners provide anecdotal evidence of the rehabilitation success of consumer support interventions (Dincin, 1975; Glasscote, Gudeman, & Elpers, 1971; Grob, 1970; Wechsler, 1960). Some research indicates that persons who become involved in social rehabilitation clubhouse programs (albeit professionally sponsored) are hospitalized less often than persons who are less involved (Beard et al., 1978; Webb, 1976; Wolkon et al., 1971; Wolkon & Tanaka, 1966). Alternative residences and support/drop-in groups that are exclusively patient-run provide an opportunity for self-determination, allowing the residents or members to control their own lives. These patient-controlled alternatives allow the helping relationship to be one of equals (Zinman, 1982).

Rappaport and associates (1985) have embarked on a longi-tudinal evaluation of GROW groups. According to Anthony and Blanch (1989), this study is the first outcome evaluation of a mental health self-help organization. The full results of this study are not yet available. However, initial results indicate that people who have been actively participating in GROW groups for more than nine months differ significantly in size of social networks, rate of employment, and measures of psychopathology (Stein, 1984) from those who have been participating for fewer than three months. Attendance at GROW meetings has been shown to be significantly related to decreases in negative coping responses such as isolation and brooding, and help-seeking responses at GROW meetings are significantly related to decreases in coping responses that rely on distraction (Reischl & Rappaport, 1988).

An article by Chamberlin (1984) has traced the growth of the consumer movement, particularly with respect to the self-help/advocacy initiatives. The consumer (or ex-patient, or ex-in-mate) movement has advocated the development of user-controlled alternatives to current mental health programs. As described by Chamberlin (1984, p. 58) these alternatives promote competence and changes.

> People who become involved in these groups may have limited perceptions of themselves; but in an environment which maximizes growth and change, they are enabled to become part of the decision-making process and are pre-vented from sinking into passive dependency. Although members may approach these organizations and services with the same limited views of themselves that they devel-oped in traditional psychiatric services (and which were reinforced there), it is only in client-controlled alternatives that they are presented with positive role models who are themselves former patients. . . . The key is choice. For-mer patients must be able to choose where to live and

what kind of living arrangement they want. People who have lost independent living skills need to relearn them, ideally from successful ex-patients who can serve as valuable role models. People who want some kind of continued care and supervision need to be able to visit a variety of living situations and pick the one they find most hospitable.

Noteworthy in the preceding quote are words consistent with the psychiatric rehabilitation approach: competence, change, choice, relearn, supervision. The actual difference between psychiatric rehabilitation and self-help is not the values or goals but the background of the helper. Proponents of self-help believe it is critical that the leadership and the helpers are consumers. They also look for persons with expertise as a source of assistance and help. As elaborated further by Chamberlin (1989, pp. 215–216):

> There are a number of similarities between the psychiatric rehabilitation approach and that of ex-patient self-help groups (given, of course, the overriding difference that the psychiatric rehabilitation approach is both professional and formal). Unlike many mental health interventions, psychiatric rehabilitation depends on the cooperation and the active involvement of the client. The stress that psychiatric rehabilitation professionals put on doing things only *with* clients (and not to or for them) is one that strikes a responsive chord in people who, all too often, have been subjected to paternalistic, coercive, or compulsory interventions. Similarly, the stress on the individualization of the psychiatric rehabilitation approach compares favorably with impersonal or depersonalizing practices that are all too familiar to most clients. For these reasons, many individuals may choose to combine being members of ex-patient/self-help groups with being clients of psychiatric rehabilitation professionals, and such a combination should work well.

Another role of the consumer movement is political action and advocacy. Local, regional, and national consumer organizations convene annual conferences and routinely speak out on issues such as informed consent, commitment laws, mental health system funding, and policy (e.g. National Mental Health Consumers Association, National Alliance of Psychiatric Survivors, NAMI Client Council). This is indeed a healthy sign for the development of both consumer-run and professionally run psychiatric rehabilitation programs tailored to the needs and wants of the consumer. When the consumer's voice is more apparent, the chances are increased that professionals will hear and listen. Without input from the consumers, programs begin to stagnate. In contrast to other fields that listen too much to the developer of the theory or the professors who teach and write about it, the psychiatric rehabilitation field must stay in touch with the needs and wants of its consumers. The primary source of learning is not books or theories but consumers and their families. The growth of both the consumer and family self-help/advocacy movement has made it easier to stay in touch.

Concluding Comment

However, the professionals' willingness to listen to consumers is sometimes less than enthusiastic. The consumer movement is like a large, powerful train coming down the tracks—the so-called Self-Help/Advocacy Special Express—with no reverse gear or brake. If we professionals do not stop, look, and listen, the train will either roll right over us or pass us by. We must find a train and a track going in a similar direction and get on that track, without trying to hook up to the Special Express because that would only slow its momentum. We can learn as we go, removing obstacles from the tracks when asked and, we hope, not recreate any new obstacles in the process.

The strength of the psychiatric rehabilitation field must remain its desire to learn from the people it is trying to help. We need to recommit ourselves constantly to listen to their voices, no matter how painful to us, at times, the message might be.

All personnel—professionals, family members, and consumers—can be more helpful if they possess the skills, knowledge, and attitudes needed to carry out their functions. Psychiatric rehabilitation attempts to empower all people, including professionals, family members, and consumers, with the skills and resources they need to do the job.

147

Programs

9

Constancy is the foundation of virtues.

Sir Francis Bacon

More and more programs describe themselves as rehabilitation oriented. These programs include community residential alternatives, community mental health centers, psychosocial rehabilitation centers, sheltered and supported work sites, and inpatient programs. Yet despite the characterization of the setting as rehabilitation oriented, it is often not clear whether the setting is actually a psychiatric rehabilitation setting, simply a variation of a traditional treatment program now offered in a community setting, or the same treatment program with a new name. For example, some programs call themselves rehabilitation programs because they serve clients with long-term psychiatric disabilities (e.g., aftercare programs, outpatient programs). Some programs call themselves rehabilitation programs because they conduct group therapy sessions that focus on functioning (e.g., day treatment centers with communication groups), provide activities in which clients interact (e.g., crafts, socialization

Parts of this chapter are excerpted with permission from:

Farkas, M. D., Anthony, W. A., & Cohen, M. R. (1989). Psychiatric rehabilitation: The approach and its programs. In M. D. Farkas & W. A. Anthony (Eds.), *Psychiatric rehabilitation programs: Putting theory into practice* (pp. 1–27). Baltimore: Johns Hopkins University Press.

classes), or provide clients with intensive support to keep them from being hospitalized. Although the setting may be different and/or the names may be different, such programs are not really psychiatric rehabilitation programs because they do not have a consistent rehabilitation orientation.

The confusion surrounding what constitutes the essential ingredients of a rehabilitation program has led to specification of the observable elements of a psychiatric rehabilitation program (Anthony, Cohen, & Farkas, 1982; Farkas, Cohen, & Nemec, 1988; Farkas, Anthony, & Cohen, 1989). No matter where the program is situated, a psychiatric rehabilitation program consists of three fundamental elements: a rehabilitation mission, a structure that promotes the rehabilitation process, and rehabilitation environments.

Mission

Many programs have mission statements, but these are often seen as bureaucratic necessities and are only referred to when preparing for official site visits or evaluation reviews. In contrast, the mission statement of a psychiatric rehabilitation program gives overall direction to the activities of the program, providing the focus for the program's activities, the criteria for program evaluation, and the rationale for revising the program.

The psychiatric rehabilitation mission is to help persons with psychiatric disabilities increase their functioning so that they are successful and satisfied in the environments of their choice with the least amount of ongoing professional intervention (Anthony, 1979). Table 9–1 presents 8 rehabilitation values reflected within this mission statement (Farkas, Anthony, & Cohen, 1989).

The first key value reflected in the mission statement is that program activities are designed to improve *functioning*. The focus of rehabilitation is on improving functioning rather than on reducing symptoms or increasing insights. The psychiatric rehabilitation program activities, therefore, focus on interventions that develop positive behaviors as opposed to interventions that control negative behavior. Interventions that improve a person's repertoire of skills and environmental supports are the predominant means

TABLE 9–1 *Key Rehabilitation Values Reflected in Mission Statement*

Functioning:	A focus on performance of everyday activities.
Success:	A focus on meeting the requirements of other people in the client's world.
Satisfaction:	A focus on the client's feelings of happiness.
Environmental specificity:	A focus on the specific context of where a person lives, learns, socializes, or works.
Choice:	A focus on self-determined goals.
Outcome orientation:	A focus on evaluating rehabilitation in terms of the impact on client outcomes.
Support:	A focus on providing assistance for as long as it is needed and wanted.
Growth potential:	A focus on the improvement in the client's functioning and status.

Adapted from: Farkas, M. D., Anthony, W. A., & Cohen, M. R. (1989). Psychiatric rehabilitation: the approach and its programs. In M. D. Farkas & W. A. Anthony (Eds.), *Psychiatric rehabilitation programs: Putting theory into practice* (p. 8). Baltimore: Johns Hopkins University Press.

for increasing functioning. This first value also implies that a psychiatric rehabilitation program is not focused on a cure or on achieving the same level of functioning for everyone. The focus is on improved functioning compared to the previous functioning of the particular individual.

The second value is *success*. Success in rehabilitation is defined in terms of meeting the demands of the environment. For example, a person who chooses to live with roommates in an apartment may need to get along with the roommates, participate in some housekeeping chores, and pay a share of the rent in order to maintain the living situation.

Although meeting the demands of the other people in a chosen environment is important, client satisfaction is critical. If the client is not happy in the environment, then the client will not want to stay there. The third value is the client's *satisfaction* with the environment.

The fourth value is *environmental specificity*. People with or without disabilities tend to function differently in different environments. For example, a person may be able to negotiate differences at work but be unable to negotiate differences with relatives at home. Psychiatric rehabilitation focuses on assessing persons in relationship to the demands of the particular environments they have chosen. Psychiatric rehabilitation programs are concerned with helping people improve their functioning in a particular living, learning, social, and/or working environment.

The fifth value is *choice*. The clients in a psychiatric rehabilitation program are seen as having the right to choose rather than be placed in an environment. Aside from the inherent decency of valuing self-determination, it is practical to have clients openly choose where they live, learn, socialize, and/or work. Whether the choice is explicit or implicit, clients, like all of us, only work to stay where they want to be.

The sixth value reflected in the mission statement is an *outcome orientation*. A psychiatric rehabilitation program is oriented toward an observable outcome rather than simply toward the provision of service. Its goal is not to provide clients with counseling and support services but rather with increased functioning and satisfaction in an environment of choice. An evaluation measure for the program that merely provides services would be to count the number of hours of services that are provided. In contrast, an evaluation measure for a psychiatric rehabilitation program is the level of functioning and satisfaction of clients in the environments of their choice.

The seventh value is the provision of *support*. Most programs view themselves as offering assistance to their clients. In a psychiatric rehabilitation program, the assistance is given for as long as it is needed and wanted. Support is not to be confused with the concept of pathological dependence. In many traditional programs, dependence is often viewed as a function of the client's illness, whereas independence is seen as doing it on one's own. The belief is that healthy people function alone and sick people are dependent on others. Yet it is clear from experience that persons with psychiatric disabilities are like all people: when given support in one area, they can often increase their ability to function in another. For example, a client who has increased support at home may be more able to arrive at work on time. In addition, the degree of support offered is a function of how much support a person wants to have. Most people have a preference as to the intensity and the length of time during which they experience assistance as supportive.

The last value reflected in the rehabilitation mission statement is belief in people's *potential for growth*. The aim of a psychiatric rehabilitation program is more than to maintain people at the same level of functioning. This is not to deny the importance of helping people stabilize themselves. People often need to concentrate on sustaining a particular level of functioning for a period of time, before growth is seen as a desirable next step. Maintenance, however, is not a sufficient aim. The mission of rehabilitation is to help people *improve* their functioning, while reducing the amount of ongoing professional intervention. Consistent with the other values, growth is seen as increasing reliance on natural support systems, improving the capacity to perform everyday activities, and decreasing reliance on professionals.

Structure

The process in a psychiatric rehabilitation program aims to achieve the rehabilitation mission. The program is structured around the rehabilitation process of diagnosis, planning, and intervening discussed in chapter 5. Three ways of structuring the program to support the rehabilitation process are the rehabilitation operating guidelines, rehabilitation activities, and the documentation of rehabilitation (Center for Psychiatric Rehabilitation, 1989; Farkas, Anthony, & Cohen, 1989). The *rehabilitation operating guidelines* consist of program policies and procedures that describe how the rehabilitation service will be provided. *Rehabilitation activities* are the organized interactions between various people in the program occurring during day-to-day program operation. *Rehabilitation documentation* is the way of recording the delivery of rehabilitation services and includes the collection of all information relevant to rehabilitation in the client records. In a psychiatric rehabilitation program, each of these three structures supports the delivery of the three phases of the rehabilitation process: diagnosis, planning, and intervening.

The process of psychiatric rehabilitation has been simply described as helping clients clarify the place they want to live, learn, socialize, or work within 6 to 24 months; assess the presence of the skills and supports needed to be successful and satisfied in that place; use the assessment information to plan for the development of skills and supports; learn the skills they lack and improve their use of existing skills; and link with resources that will give them the support they need and modify resources to improve their supportiveness (Anthony, Cohen, & Cohen, 1983).

Psychiatric rehabilitation programs revolve around providing opportunities for a rehabilitation diagnosis, rehabilitation plan, and rehabilitation interventions to occur. The program's operating guidelines, activities, and documentation all work together to provide clients with an opportunity to choose where they want to live, learn, socialize, or work and then to assess and develop the skills and supports they need to be successful and satisfied in that environment. Because the psychiatric rehabilitation process is extensively described in chapters 5, 6, and 7, only some of the implications of the process for designing a program are discussed here.

Diagnosis in Rehabilitation Programs

Diagnostic Operating Guidelines. The program's operating guidelines require that a rehabilitation diagnosis be conducted for each client. For example, a policy statement might state that clients should be assessed in terms of their readiness to participate in choosing an overall rehabilitation goal; that clients should receive whatever assistance they need in developing their ability to participate in choosing a goal; that a functional and resource assessment be conducted for all clients; and that the program be organized to maximize client participation in the rehabilitation diagnosis. The procedures might detail who should perform the rehabilitation diagnostic tasks and how they are accountable for the fullest client involvement possible. For instance, in order to establish the clients' readiness to set a rehabilitation goal, the intake worker or the program's gatekeeper might explore the clients' degree of dissatis-faction with their current environments, as well as the level of dissatisfaction of persons within the current environments. This same person might also go on to assess the clients' knowledge of their own values and of different environments.

Diagnostic Activities. The diagnostic activities revolve around the diagnostic tasks specified in the technology for rehabili-tation diagnosis (M. R. Cohen et al., 1986, 1990). For example, to set an overall rehabilitation goal, the client and the practitioner must first explore the client's choice of an environment. This program activity might begin with an assessment of interests by a professional psychometrician (e.g., vocational interest testing or academic interest testing) followed by face-to-face interviews with the rehabilitation practitioner to translate the assessment infor-mation into personal criteria for making a choice. The second program activity involves researching different possibilities so that the client becomes informed about alternative environments. Third, the practitioner leads the client through a decision-making method that uses the information gathered by the client to evaluate the alternative environments. People the client feels are relevant to the decision share their points of view. The goal is then set.

The goal-setting process implies that some activities are best scheduled in an office or group room. The client and the practitioner can work alone or in a group to clarify values, analyze past experiences, and complete a problem-solving matrix. Other goal-setting activities need to be completed by traveling to different settings or meeting with various people (e.g., researching possible environments, gathering information from relevant others). When the program's policies and procedures require such goal setting, an adequate amount of time is allocated for each activity, and the client/staff ratio (15:1) is low enough to allow practitioners the time to engage in the goal-setting activities. In a recent review of residential, educational, and vocational programs that use the psychiatric rehabilitation approach, the amount of time set aside for conducting a rehabilitation diagnosis varied from 2 to 3 sessions during a 2-week period in a vocational setting (Brown & Basel, 1989) to 12 sessions during 4 weeks to set an overall rehabilitation goal in a hospital setting (Lang & Rio, 1989). An educational rehabilitation program set aside 72 sessions during 8 months to conduct a functional assessment (Hutchinson, Kohn, & Unger, 1989), while a community supervised apartment program designed their functional assessment to take 2 weeks (Mynks & Graham, 1989). Clearly, even if program policies concerning the process of rehabilitation diagnosis sound remarkably similar in varying types of programs, the activities can be expanded or contracted to fit the time frames and staffing patterns of different program structures across the gamut of living, learning, social, and working programs, both in the hospital and in the community.

Diagnostic Documentation. The records reflect the importance of setting an overall rehabilitation goal by calling for documentation of the steps of the process and the statement of a client's final choices. The actual worksheets used during the process might be filed elsewhere to keep the client record manageable. The design of the client records would require an evaluation of the client's skills and supports for each individual client in relation to each specific environment they have chosen. An example of the assessment information that might be in the records appears in Tables 6–1 and 6–2 in chapter 6.

Planning in Rehabilitation Programs

Planning Operating Guidelines. The planning phase uses the diagnostic information to select and plan for the development of high-priority skills or supports. Each skill or support selected for development is specified as an objective in the rehabilitation plan. Each objective is assigned a specific intervention with specific persons responsible for delivering the intervention in a specific time period. The program's policies state that clients will participate in meetings to develop their plans once a rehabilitation diagnosis has been completed. Procedural statements specify the personnel responsible for each planning task, how the priorities for the objectives in the plan are set, the personnel responsible for monitoring the implementation of the plan, and the personnel responsible for revising the plan in the event of insufficient progress in achieving the objectives.

Planning Activities. The design of planning activities includes sufficient time for maximizing the involvement of the client and significant others in developing a plan. The activities include preparation meetings with the client and significant others as well as a meeting of all parties to discuss the plan. In practice, such a planning process is usually conducted with staff from a variety of disciplines. In some programs, the person who coordinates the planning process is the case manager (Goering, Huddart, Wasylenki, & Ballantyne, 1989); in other programs, a mental health attendant (Craig, Peer, & Ross, 1989).

Planning Documentation. The documentation of the plan includes a statement of the overall rehabilitation goal, the high-priority objectives, the interventions to address each objective, the persons responsible for different tasks, and the signatures of the client, practitioner, and significant others who participated in planning and agree with the plan. Table 7–1 in chapter 7 presents an example of a rehabilitation plan.

Intervening in Rehabilitation Programs

Intervention Operating Guidelines. In the intervening phase, the plan is implemented by either skill development or resource development interventions. Skill development interventions directly teach clients the skills they lack (direct skills teaching) or develop their use of the skills they do not use effectively (skills programming). If the plan calls for resource development, the interventions used either link clients with existing resources (resource coordination) or modify existing resources to improve their supportiveness (resource modification).

An example of a policy statement related to intervening might be that interventions will be conducted, as much as possible, in the environment specified in the client's overall rehabilitation goal. The procedures specify the tasks of the practitioners, clients, and significant others during the intervention in the specified environment and might include the steps the practitioner would follow to help people in the environment support the client's use of a newly learned skill. For example, the procedure might specify that residential skills are learned at the learning center, and then residential counselors monitor the client's use of the skills in the residence (Rice, Seibold, & Taylor, 1989).

Intervention Activities. The program activities are designed to ensure that skill development and resource development interventions are conducted. The daily schedule of the program may be designed with blocks of time for individual or group skills teaching sessions. Similarly, classes may be scheduled for a skill area frequently needed by all clients. The process of linking clients with resources involves having staff time and equipment (e.g., cars, vans) to transport clients to visit alternative resources and to provide follow-up assistance once the client is receiving support from a resource.

Intervention Documentation. The documentation of the rehabilitation interventions is most similar to existing documentation procedures. The dates of provision of the intervention are recorded. In addition, copies of the lesson plans used for skills teaching and the programs developed during skills programming are referenced. Contacts with resources are also logged. Most important is the recording of progress, whether it is improved skill use or improved support.

A psychiatric rehabilitation program organizes its operating guidelines, activities, and documentation so that ongoing program evaluation can be done. Evaluation information is collected on the program's consistency with the values implied in the mission statement (e.g., the number of clients in environments they have selected, the number of days spent in those environments, or the degree of client success and satisfaction in those environments). Table 9–2 presents program evaluation measures for quality control and quality assurance. The outcomes might suggest a need for changes in how the program process occurs or suggestions for service system change. For example, if the types of housing that clients preferred were unavailable, then the program might wish to join with other programs to suggest to the mental health authority that housing policies be changed. If the outcome is, however, that clients do not remain successful and satisfied in the settings they chose, the implications may be that the program guidelines, activities, or documentation needs to change, or, alternatively, that the practitioners are not delivering the process as well as they might and that some change is required in their knowledge, attitudes, and/or skills. Last, the results may imply that the service system needs to alter its policies, for example, to require staff support of clients' choices of places to live, learn, socialize, and work.

TABLE 9–2 *Program Evaluation*

Quality Control

Rehabilitation Mission	*Examples of Mission-Related Outcomes*
Increase functioning so that people are successful and satisfied in their environments of choice with the least amount of ongoing professional intervention.	Percentage of clients who stay in environments they have chosen. Number of days/month clients stay in environments they have chosen as compared to previous period of time. Percentage of clients who are satisfied in environments they have chosen. Percentage of clients who are performing daily activities with less professional intervention.

Quality Assurance

Rehabilitation Process	*Examples of Process-Related Outcomes*
Diagnosis/planning/intervention	Evidence that clients choose their environments. Evidence that clients are evaluated in terms of what they need to do or have to obtain and/or stay in environments of their choice. Evidence that all program activities result in clients doing or getting what they need to be successful and satisfied in their environments. Evidence that all program activities are designed around the needs and preferences of program clients. Evidence that all program activities involve clients as partners.

In summary, the operating guidelines, program activities, and documentation in a psychiatric rehabilitation program ensure that the psychiatric rehabilitation process is embedded in the program structure. The entire process of psychiatric rehabilitation occurs within one or many environments. The next section describes the environments of a psychiatric rehabilitation program.

Program Environments

During the past 15–20 years, rehabilitation programs were often defined by *where* the program settings were located. In the ongoing debate about whether community-based programs were better than hospital-based programs, some observers felt that a rehabilitation environment could not exist within a hospital setting. Rehabilitation was deemed, by definition, to occur only in community-based settings. No matter how innovative a hospital environment was, no matter how little a community-based program represented rehabilitation values, the program located in the community was the only one viewed as a rehabilitation program.

The philosophy and technology of psychiatric rehabilitation are certainly community focused. People, with or without disabilities, generally wish to live, learn, socialize, and work not in artificial settings but rather in the real world of the community. Very few people who are psychiatrically disabled prefer to live for a long period of time in an institution (Center for Psychiatric Rehabilitation Staff, 1989). Yet even though psychiatric rehabilitation is always community focused, it does not always need to be community based.

Several reasons are usually cited for the continued existence of psychiatric institutions. First, inpatient settings provide long-term shelter and security (Bachrach, 1976). Second, the political structure of state and local communities often find institutions easier to fund, organize, and evaluate than a series of community-based settings with a diversity of responsibilities, geography, and organizational structures. Third, some clients prefer to live in institutions, if they are allowed to, for a short time when they feel the need for stabilization. Such stabilization may also occur in a community environment, a widely known but little appreciated fact, at least at present. For these reasons, for better or worse, psychiatric institutions are likely to continue to exist for the foreseeable future.

As long as they do exist, it makes sense not to ignore the role they can play in the field of mental health, but rather to encourage them to be guided by a rehabilitation philosophy, that is, devoted to helping persons function better in the community of their choice. Both inpatient and community settings should be judged in terms of how well they aim toward this mission.

As highlighted in this chapter, the psychiatric rehabilitation process shapes the program structure. Similarly, the psychiatric rehabilitation philosophy shapes the rehabilitation environments, defined by its network of settings and the context in which the psychiatric rehabilitation program operates (Center for Psychiatric Rehabilitation, 1989; Farkas, Anthony, & Cohen, 1989).

Network of Settings

The settings surrounding the program over which the program has direct control is called a *network*. Settings the program can use, link with, or refer clients to are not part of the program's network of settings, but are part of the overall service system in which the program operates. Thus, a program may hold the lease for a number of apartments in a geographic area. These apartments are within the program's control and form part of the program's network of settings. A supported employment program that provides support for clients holding competitive jobs would link clients with job sites but would not administer or have authority over the job sites. The jobs themselves, therefore, are part of the overall service system in which the program operates, but are not part of the program's network of settings. The program's settings are usually linked to the larger service systems by interagency agreements, individual liaisons, or case managers.

The settings for the program can be residential, educational, social, or vocational settings. Although people may socialize in every environment, sometimes one specific setting has socializing as its primary purpose (e.g., a club or social organization). The settings can be managed by the mental health authority (such as medication clinics or group homes) or be natural settings, such as apartments, competitive jobs, or university programs.

Psychiatric rehabilitation settings accommodate different client preferences. Ideally, programs include a variety of settings that match client preferences. For example, a vocational rehabilitation program may need to expand its settings to include the types of places its current group of clients might like to work, rather

161

than having a fixed series of work activities or work sites that fit for the previous group of clients. Some clients like manual labor such as janitorial service, landscaping, or woodworking—jobs that are common in many rehabilitation programs. Other clients, however, would prefer to work as accountants, computer programmers, or counselors. Some settings might have mental health supervision, and others might be natural settings not specifically designated for clients. An educational program may offer instruction in a day treatment setting or in adult education courses in a high school, or it may provide support for individuals who choose to go to a local university. Preference would also dictate the location of the programs' settings. A psychiatric rehabilitation program asks the clients in its program about their preferences. Programs that organize their settings by preference include mechanisms that allow for altering the settings to fit the preferences of the current clients.

In addition to preference, a psychiatric rehabilitation program accommodates clients' levels of functioning. If a program has an array of residential, vocational, social, or educational settings but the entry criteria are higher than the current client populations' level of functioning, then clients' needs are not being met. Some client groups may, in fact, cluster together in terms of their current abilities. As a result, the program should offer a variety of settings with the same functional demands. The network of settings may be integrated by having the functional exit demands of one setting match the functional entry demands of another setting. The program process then is organized to help the client move from the entry-level demands to those of the exit criteria. The difference between an integrated network of settings and a continuum of care is that in an integrated network of settings clients are not required to move from one functional level to another simply because it is the next level up. A client in a program with an integrated network of settings may be very successful in one setting and choose to remain there. Also, as the client's functional level improves, the client may not be required to change settings at all. Rather, the level of support within the setting might change. This feature is very common in rehabilitation programs labeled as supported work, supported housing, or supported education (Farkas & Anthony, 1989).

Context

The context of the environment refers to the cultural or organizational beliefs reflected in the arrangement of the physical decor, the types of activities offered that are not focused on the rehabilitation process, and the administrative practices. A rehabilitation environment reflects rehabilitation values in each area: physical layout, generic activities, and administrative practices. For example, the value of choice implies that the space reflects the taste of the clients.

A rehabilitation environment often has many activities not designed to diagnose, plan, or intervene with the clients in environments they prefer. Some programs have traditions about observing staff and client birthdays or important personal events. Some programs encourage staff and clients to be friends and to see each other in leisure-time activities; others do not. Some programs offer sports activities for client enjoyment, and some are actively engaged in community education. In a psychiatric rehabilitation program, these other activities also occur within the framework of rehabilitation values.

Last, the context of a rehabilitation environment includes the administrative practices of the program. An administration that values client involvement might arrange its program hours of operation to reflect greatest client need rather than staff convenience. For example, the program may be open evenings and weekends, rather than 9 A.M. to 5 P.M. every day. Client involvement might also be reflected in the degree to which consumers or ex-patients participate in designing and operating the program. An advisory board with client members who do not have the skills or supports to make themselves heard is not a reflection of a rehabilitation environment that values choice; rather, a rehabilitation environment provides clients who wish to participate on the board with the skills and supports to do so effectively. A growth- and outcome-oriented program treats staff in a way consistent with these values. Staff are given the opportunity to develop their knowledge and skills, and staff effectiveness is measured in terms of the outcomes they produce.

In summary, the context of the environment is consistent with rehabilitation values. Table 9–3 presents a summary of the

TABLE 9–3 *Basic Psychiatric Rehabilitation*
Programming Principles

Mission

1. A program is organized and evaluated around its mission statement.

Structure

1. A program provides opportunities and assistance in choosing, getting, and keeping environments.

2. A program maximizes the degree to which clients are involved in the process of their rehabilitation.

3. A program involves clients in selecting the environments they prefer rather than matching clients to environments based on their levels of functioning.

4. A program provides skill and support development activities specific to individual clients.

5. A program helps clients play an active role in developing their own supports.

6. Evaluation of the program's achievement of client goals provides direction for program growth and change.

Environment

1. A program includes a range of settings that reflect client preferences and levels of functioning.

2. Programs locate themselves in or resemble naturally occurring environments in which people live, learn, work, or socialize.

3. A program's decor, generic activities, and administrative practices are consistent with rehabilitation values such as involvement, choice, functioning, outcome orientation, growth potential, support, and satisfaction.

important principles inherent in a rehabilitation program's mission, structure, and environments.

Present Day Psychiatric Rehabilitation Settings—How Do They Measure Up?

It is possible to take these basic program elements of mission, structure, and environments and define them in greater detail. These definitions can then be used to assess psychiatric rehabilitation settings comprehensively. In this way, any setting can be assessed with respect to its current capacity to implement a rehabilitation approach.

The Boston University Center for Psychiatric Rehabilitation has been involved in projects that have assessed more than 100 different settings, both inpatient and outpatient. Typically, the methods of assessment involve a 1–3 day site visit to collect data by means of observations of the settings' activities, interviews with clients and staff, a review of sample case records, ratings of practitioner responses to standard client situations, and sometimes the submission of audiotapes of practitioners diagnosing, planning, and intervening with clients. All of these data (as well as a review of agency brochures and other written material) are then analyzed off-site in an attempt to understand the setting's strengths and weaknesses with respect to implementing the psychiatric rehabilitation approach. The analysis is organized conceptually around the dimensions of personnel, programs, and service systems. As introduced in chapter 8, psychiatric rehabilitation can be more easily understood by identifying the personnel, program, and service system variables that contribute to client rehabilitation outcome. In other words, personnel, programs, and service system variables become the units of analysis for assessing rehabilitation settings. Each of these three dimensions represents a window for investigating any setting.

The personnel unit of analysis examines the skills, knowledge, and attitudes of the practitioners in the setting. The program unit of analysis examines the capacity of the setting's mission, structure, and environments. The system unit of analysis examines whether the service system's policies and functions facilitate the capacity of the personnel and the program to conduct rehabilitation diagnosis, planning, and interventions with clients.

One assessment study that was part of a larger technical assistance project examined 54 adult partial care settings (i.e., day treatment programs) receiving state department of mental health funding in New Jersey (Fishbein, 1988). Assessment results indicated that the rehabilitation program settings were particularly deficient in:

1. Documentation of the skill deficits and especially the skill strengths of clients.
2. Documentation and the evaluation of clients' present and needed skill functioning.

3. Documentation of the strengths and deficits in the clients' supports.
4. Documentation of the relative importance or priority of different client goals.
5. Client awareness of their signing their rehabilitation plan.
6. Agency policies specifying client graduation or termination criteria.

Also noted was the wide range of agency functioning on all the different measures, both within and between agencies. In general, however, the partial care settings in this investigation did not emphasize the rehabilitation diagnostic process. The clients' more general problems or needs rather than the specific assessment of client skills and supports were recorded. Without proper rehabilitation diagnoses, the capacity of the setting to develop needed skills and supports is limited. The rehabilitation diagnoses drive the clients' rehabilitation plans and interventions and fashion the program environments. Without such rehabilitation diagnoses, the achievement of positive client rehabilitation outcomes is at best problematic and at worst, impossible.

The Center for Psychiatric Rehabilitation has reported the results from its on-site assessments conducted at hospital and community-based settings (Farkas, Cohen, & Nemec, 1988). The data generated information relevant to each setting's capacity to implement a psychiatric rehabilitation approach. The settings in this particular assessment study represent 40 agencies, involving 94 different programs that deliver services to persons with severe psychiatric disabilities. The agencies were located in 1 Canadian province and 12 American states ranging from Oregon and Ohio to Virginia and Maine. Twenty-one agencies applied to the Center to participate in a project that would train agency personnel to become trainers in psychiatric rehabilitation. To select the most compatible agencies for the project, each applicant agency was assessed to ascertain the degree of compatibility between the applicant setting and ideal psychiatric rehabilitation practice. The remaining 19 settings specifically asked the Center to conduct an assessment to assist them in improving their services for persons with severe psychiatric disabilities. Both subgroups, representing a wide variety of settings, expressed a desire to improve their psychiatric rehabilitation services.

The average number of staff for the 40 agencies was 54. Staff size ranged from 3 to 731. The average staff-to-client ratio was 1:7, and the range of ratios was wide—from 1:2 in hospital settings to 1:44 in community mental health centers.

The average client caseload of the 40 agencies was 191 clients, ranging from a low of 22 (residential facilities) to 1,132 (state hospital facility). The average age of clients was 33.6 years old with a range from 11 years of age to 90. The average number of previous hospitalizations among the clients of the 40 agencies was 3.7, ranging from 1 previous hospitalization to 16. In summary, the agencies represent the range of facilities serving the severely psychiatrically disabled population across the United States. The most staff-intensive facilities were, obviously, the inpatient settings, and psychosocial rehabilitation centers and mental health clinics were the least staff intensive.

Farkas, Cohen, and Nemec (1988) detailed the ingredients of each phase of the psychiatric rehabilitation approach selected for analysis and gave a thorough report. In general, the results suggested that the concept and methods of rehabilitation are talked about more than they are practiced. For example, although the settings indicate that skills teaching is definitely their orientation, only about 30% define the client skills being taught or provide skill training in the context of the clients' overall rehabilitation goals. Furthermore, although 75% of the settings claim to stress client involvement in the assessment and planning process, less than 40% recorded any evidence that the clients were actually involved.

In essence, many settings give lip service to the concept of rehabilitation, but few use the technology for implementing it. The promising aspect of this study is that these settings attempting to implement a rehabilitation approach actually recognize their deficiencies and are actively exploring ways to improve their use of rehabilitation technology.

167

Concluding Comment

Certainly more programs are attempting to serve persons with psychiatric disabilities, and more programs are adopting a psychiatric rehabilitation orientation. In order for these programs to impact client-rehabilitation outcome, they, like their clients, must have the necessary skills and supports. Chapters 5, 6, 7, and 8 identify many of the skills practitioners need to deliver a rehabilitation program effectively. Chapter 10 describes the functions of the service system and how the system can support the agencies' programs and personnel. Personnel, programs, and service systems operate synergistically. In order for programs to be maximally effective, they need skilled personnel and supportive service systems. Similarly, skilled personnel need to practice in programs and service systems that allow them to use their skills. Likewise, service systems need skilled personnel and excellent programs if their system functions are to positively impact the lives of persons with psychiatric disabilities.

10

Service Systems

> *Love our principle, order our foundation, progress our goal.*
>
> *Auguste Comte*

From a psychiatric rehabilitation perspective, the most basic mission of the service system is to increase the chances that persons with psychiatric disabilities will be helped by expert personnel in state-of-the-art psychiatric rehabilitation programs. In essence, the primary purpose of the service system is to use the most effective personnel in the most effective programs to ensure that persons who are psychiatrically disabled achieve their rehabilitation missions. In other words, the skilled personnel who help persons with psychiatric

---

Parts of this chapter are excerpted with permission from:

Anthony, W. A., & Blanch, A. (1989). Research on community support services: What have we learned? *Psychosocial Rehabilitation Journal, 12*(3), 55–81.

Cohen, M. (1989). Integrating psychiatric rehabilitation into mental health systems. In M. D. Farkas & W. A. Anthony (Eds.), *Psychiatric rehabilitation programs: Putting theory into practice* (pp. 162–191). Baltimore: Johns Hopkins University Press.

Cohen, M., & Anthony, W. A. (1988). A commentary on planning a service system for persons who are severely mentally ill: Avoiding the pitfalls of the past. *Psychosocial Rehabilitation Journal, 12*(1), 69–72.

disabilities are supported in conducting rehabilitation activities by effective programs; these programs and personnel are in turn supported by effective service systems.

The technology used by personnel involved in psychiatric rehabilitation is described in chapter 5. The program mission, structure, and network of environments are described in chapter 9. This chapter suggests how the philosophy, policies, and administrative functions of the mental health service system can support the integration of psychiatric rehabilitation personnel and programs into mental health service systems.

The Need for System Support for the Psychiatric Rehabilitation Approach

Perhaps the most straightforward definition in general—and a definition most relevant to the mental health service system in particular—is that a service system is a combination of services organized to meet the needs of a particular population (Sauber, 1983). As is well known, persons with long-term psychiatric problems have multiple residential, vocational, social, and educational needs. Between 1800 and 1950, these multiple needs typically were met within state institutions (originally called insane asylums, then state hospitals, and now mental health or psychiatric institutes or centers). State institutions functioned as settings in which persons with mental illness could be taken care of during their illness or, if necessary, throughout their lives. Although active treatment was emphasized, the de facto mental health mission for this population was custody. Rehabilitation was offered as an ancillary service within the hospital, seen primarily as providing activities. In many respects, the mental health service system was organized to support the state institution, not the individual client.

With the discovery of psychotropic medication, chemotherapy became a preferred mode of treating persons with long-term psychiatric disabilities within institutions. The use of chemotherapy to reduce symptomatology, along with a number of other factors including changing federal reimbursement policies, raised consciousness about patients' rights and the high costs of institutional care, helped make the social reform of deinstitutionalization possible (Brown, 1982; Rose, 1979; Williams, Bellis, & Wellington, 1980). By the late 1970s the resident population of state psychiatric hospitals had dramatically declined (Bassuk & Gerson, 1978).

Deinstitutionalization has radically changed how persons with severe mental illness are served by the mental health system. As discussed in previous chapters, many persons with prolonged mental illness are now receiving chemotherapy while functioning in residential, vocational, educational, and social settings in the community. Many other persons with psychiatric disabilities have rejected these settings and are not engaged in mental health services at all. In general, neither the state hospital-based system nor the community-based system has been very successful in helping persons with severe psychiatric disabilities achieve their rehabilitation goals (Anthony et al., 1972, 1978; Anthony & Nemec, 1984).

In addition, the change in location of services from the hospital to the community has created decentralized, diverse services. The diversity of services complicates their organization in the mental health system. Previously, the mental health system was organized around the state hospital, but the current diversity of services requires an interdependent organization of multiple community services (Gittleman, 1974).

Another difficulty in creating a mental health service system for persons with psychiatric disabilities stems from the varied, multiple needs of the client population (Scott & Black, 1986). Many different service systems could be designated as responsible for meeting the individual needs of persons with long-term psychiatric disabilities (e.g., vocational rehabilitation, social security). The diverse needs of persons with severe psychiatric disabilities for housing, health care, economic, educational, vocational, and social supports dictates coordination between many existing service systems. The mental health service system, however, is the primary system responsible for preventing individuals who need services from being ignored or falling through the cracks. The challenge is to develop a mental health service system that can consistently meet the diverse needs of all clients (Reinke & Greenley, 1986). In essence, not only must effective and relevant services be available, but they must also be well-coordinated so that they are easily accessible and efficient.

Research on Service Systems

Although many studies have noted that multiple, fragmented service systems can interfere with effective service delivery to persons with psychiatric disabilities, little systems-level research has been undertaken (Anthony & Blanch, 1989). In 1977, Armstrong reported on 135 federal programs in 11 major departments and agencies that had direct impact on the mentally ill. He reported that many of the failures of deinstitutionalization could be attributed to funding disincentives and lack of coordination among these programs. Other evidence of the need for system development and integration includes the interrelationship of health and mental health as demonstrated by the frequent conflict between services rendered by primary care physicians and mental health professionals (Burns, Burke, & Kessler, 1981); existing funding streams with conflicting regulations and eligibility criteria (Dickey & Goldman, 1986); and data indicating that consumers and family members know less about existing services than providers (Grusky & Tierney, 1989). Moreover, the lack of coordination directly affects clients. Tessler (1987) found that when clients do not connect with resources after discharge from inpatient care, their overall community adjustment is poorer and there are more complaints about them.

On the other hand, poor coordination is sometimes blamed for failures actually due to insufficient resources or inappropriate services (Solomon, Gordon, & Davis, 1986). Research has not yet clarified the relationship between the numbers, types, or coordination of services and client outcome.

Many efforts have been made over the years to address the issue of service coordination. The community mental health center (CMHC) movement itself was an attempt to provide an array of interlinked services within a single organization and to assure availability of services through a catchment area. A second example is the National Institute of Mental Health (NIMH) linkage initiative in the late 1970s that funded liaison positions between CMHCs and primary health care projects. Each program experienced some successes, particularly in increasing the percentage of the population receiving mental health services. They were

172

less successful, however, in coordinating services for people with severe psychiatric disabilities (Dowell & Ciarlo, 1983; Goldman, Burns, & Burke, 1980; Tischler, Henisz, Myers, & Garrison, 1972).

Anthony and Blanch (1989) categorized various attempts at ensuring the integration of services into four types according to whether they emphasized (1) legislated relationships and program models, (2) financing mechanisms, (3) strategies for improving interagency linkages, and/or (4) assignment of responsibility. Many initiatives have, of course, incorporated several of these elements.

Legislated Relationships and Program Models

Georgia's balanced service system model, New York's unified services legislation, and California's model program standards were early attempts to legislate relationships between state, county, and local providers and to describe and fund a specific set of services. Case studies of these initiatives reveal mixed success and failure. In New York, the program was optional and was not widely disseminated. It was quite successful, however, in five counties that did participate (Pepper & Ryglewicz, 1982, 1983). In California, the budgetary implications of the proposed model were unacceptable to governmental decision makers (Barter, 1983), and in Georgia, changes in political leadership mitigated the original promise of the model (Gay, 1983; Miles, 1983).

Several attempts have also been made to evaluate the introduction of community support programming through state legislation and funding. A historical analysis of hospitalization rates in Oregon (Hammaker, 1983) shows a period of backsliding and lack of coordination of services in the late 1970s, no real changes during a period of statewide community support program (CSP) planning (1977–1979), and a dramatic decrease in hospital bed/day use when funding and monitoring of CSP services actually began (1980–1982). Similarly, Lannon, Banks, and Morrissey (1988) demonstrated improvement or maintenance of high levels of community tenure for older CSS (community support system) clients in New York state, although there was no improvement for younger clients.

Financing Mechanisms

Recently, attempts have been made to improve service integration through new financing mechanisms. Many of these initiatives build on the notion of centralized clinical and fiscal responsibility in the same administrative structure, a concept that has worked well in Dane County, Wisconsin (Dickey & Goldman, 1986). For example, the Robert Wood Johnson Foundation has funded several pilot projects that pool existing funds through a single mental health authority (Rubin, 1987). Similar experiments are being tried with Medicaid and Medicare demonstration sites, health maintenance organizations, and regional authorities for comprehensive care (Dickey & Goldman, 1986). No final reports are yet available on the impact of these programs on service use or client outcomes.

Interagency Linkages

Pincus (1980) describes six different models for interagency linkage based on contractual, functional, and educational relationships. Other researchers have discussed the issues involved in facilitating cooperative efforts between different services organizations (e.g., Woy & Dellario, 1985). Empirical research, however, is scant. Dellario (1985) found a trend towards improved vocational outcomes for clients served by mental health and vocational rehabilitation agencies with good interagency relationships, but the trend is not significant. Similarly, Rogers, Anthony, and Danley (1989) found improved vocational outcomes in two pilot areas in Vermont participating in interagency training and joint policy-making activities; other areas in the state did not show the same increase until 2 years later. Several case studies also describe different ways of organizing community support systems to facilitate interagency cooperation, but no outcome data are available (Grusky et al., 1985; Morrissey, Tausig, & Lindsey, 1985).

Assignment of Responsibility

A fourth strategy for improving service integration (often used along with other initiatives) is to identify a specific group of clients and assign responsibility for their care and treatment

174

to an individual, team, or organization. Recent examples of this approach include the *core service agency* or *lead agency* concept, as well as various case management models that designate specific pools of high risk or high demand clients. These studies, many demonstrating positive outcomes, are reviewed in the case management section of this chapter.

Response to System Deficiencies in the 1980s

During the decade of the 1980s, there were two predominant responses to the problems caused by the lack of critical services and the fragmentation of those few services that did exist. The two initiatives were the development and implementation of the Community Support System (CSS) model, and the renewed recognition of the value of case management services.

Community Support System

In the mid-1970s a series of meetings at NIMH gave birth to the idea of a Community Support System (CSS), a concept of how services should be provided to help persons with long-term psychiatric disabilities (Turner & TenHoor, 1978). Recognizing that current services were unacceptable, the CSS described the array of services that the mental health system needed for persons with severe psychiatric disabilities (Stroul, 1989). The CSS filled the conceptual vacuum resulting from the aftermath of deinstitutionalization (Test, 1984). The CSS was defined as:

> A network of caring and responsible people committed to assisting a vulnerable population meet their needs and develop their potentials without being unnecessarily isolated or excluded from the community. (Turner & Schifren, 1979, p. 2).

The CSS concept identifies the essential components needed by a community to provide adequate services and support to persons who are psychiatrically disabled. Figure 10–1 graphically portrays these essential components.

The CSS initiative was launched in 1977 to assist states and communities in developing the comprehensive and integrated

175

FIGURE 10–1 *Community Support System*

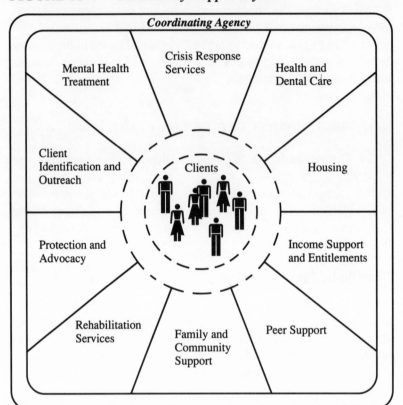

Reprinted from: Stroul, B. (1989). Community support systems for persons with long-term mental illness: A conceptual framework. *Psychosocial Rehabilitation Journal*, *12*, 9–26.

range of services that comprise a CSS. This initiative came to be known as the NIMH Community Support Program (CSP). At CSP's inception, federal grants were provided to state mental health agencies in order to help them develop CSSs in their states. Every state, the District of Columbia, and two territories have received federal funds from CSP. Technical assistance has also been provided routinely to states and communities. In 1986 CSP also provided funds for demonstration projects to develop and evaluate community-based approaches for persons with psychiatric disabilities who are elderly, or homeless, or young adults with substance abuse problems. In 1989 CSP funded research projects designed to evaluate the essential CSS services of case management, crisis intervention, and psychiatric rehabilitation.

The essential components of a CSS have been demonstrated and evaluated since its inception. As mentioned in chapter 3, Test (1984) concluded from her review that programs providing more CSS functions seem to be more effective (fewer rehospitalizations and improved social adjustment in some cases) than programs that provide fewer CSS functions. More recently, Anthony and Blanch (1989) reviewed CSS-relevant data and concluded that research in the 1980s documented the need for the array of services and supports originally posited by the 1975–1977 CSS concept. It appears that the need for CSS-component services has a base in empiricism as well as logic.

Furthermore, Anthony and Blanch (1989) concluded that the CSS research agenda is now poised for an explosion of meaningful research capable of forming policy and changing the configuration and delivery of services to persons who are psychiatrically disabled. Some data suggest the future research direction of each CSS component. Most important, interventions relevant to most CSS components can now be described at a level of detail that permits their implementation to be observed, measured, and monitored reliably. A significant number of quasi-experimental and small-scale experimental studies have been carried out that show future CSS research is not only needed but also increasingly feasible. The stage is now set for larger, long-term research studies of measurable, replicable CSS services.

Case Management

One essential service in the CSS is case management. In contrast to systemwide initiatives that attempt to build a better system of services for all persons with severe psychiatric disabilities living within a specific geographic area, case management attempts to build a better service system around each individual client, one client at a time.

Case management helps persons with long-term psychiatric disabilities obtain the services they want and need (Anthony, Cohen, Farkas, & Cohen, 1988). Services, in this instance, are not just provided by formal mental health agencies, but can be provided by any capable person, agency, or organization chosen by the client.

As noted earlier in this chapter, case management has been seen as a response to system inadequacies, particularly service rigidity, fragmentation, poor use of services, and inaccessibility (Joint Commission on Accreditation of Hospitals, 1976). However, Anthony, Cohen, Farkas, and Cohen (1988) view case management as more than just a response to a dysfunctional system. They see case management as a needed function no matter how coordinated and integrated the system. Case management is a uniquely personal response to a client's specific service needs and overall rehabilitation goals. From this perspective, case management brings to life the human dimension of the human service system (Anthony, Cohen, Farkas, & Cohen, 1988).

Case Management Models. Case management models differ widely in their philosophy, focus, and activities. The underlying values of different models vary in terms of whether they are person or illness oriented and whether they are client or service-system driven. In some models case management programs provide only service coordination; in others, crisis intervention, rehabilitation, and treatment services are also provided.

The focus of case management varies most depending on the model. Four leading models of case management were identified by the Mental Health Policy Resource Center (1988): the personal strengths model, the Program of Assertive Community Treatment (PACT) model, the broker model, and the rehabilitation case management model. The personal strengths model focuses on securing natural community resources that will support the clients' existing strengths (Mental Health Policy Resource Center, 1988; Modrcin, Rapp, & Chamberlain, 1985). The PACT model uses an assertive outreach team to provide comprehensive services to maintain clients in the community (Brekke & Test, 1987). The broker model focuses on linking clients to mental health services (Mental Health Policy Resource Center, 1988), and the rehabilitation case management model focuses on developing clients' skills and supports to increase their success and satisfaction in a chosen environment (Anthony, Cohen, & Cohen, 1983). In addition to the models identified by the Mental Health Policy Resource Center (1988) there are at least two other well-known models: the clinical case management model (Harris & Bergman, 1987a, 1988b, 1988c) which focuses on supporting clients through personal counseling and environmental modification; and the advocacy model, which focuses on empowering clients (Rose, 1988). Although the philosophy and focus vary greatly among different case management models, there is a consensus about certain core activities. Levine and Fleming (1984) identified six core case management activities: identification and outreach, assessment, planning, linking, monitoring, and advocacy. The core activities are conducted within program environments that vary in caseload size, service site, intensity of contact, duration of contact, credentials and salaries of case managers, staff configuration, supervisory structure, agency culture, program costs, and the quality of the surrounding service system. Differences in the implementation of the models, as well as the variations in the program environments and service systems in which they are used, create difficulties in comparing the outcomes of different case management models.

Case Management Outcome Studies. Most case management research has been descriptive studies of the characteristics of the program environments in which case management occurs. The research literature describes the characteristics of case managers (Intagliata & Baker, 1983; Caragonne, 1981; Goldstrom & Manderscheid, 1983), case management activities (Baker & Weiss, 1984; Berzon & Lowenstein, 1984; Caragonne, 1983; Intagliata, 1982; Kurtz, Bagarozzi, & Pollane, 1984; Levine & Fleming, 1984; Marlowe, Marlowe, & Willets, 1983), ideal caseload size (Intagliata & Baker, 1983; Schwartz, Goldman, & Churgin, 1982), and the relative merits of individual as compared to team case management (Turner & TenHoor, 1978).

In the 1980s, outcome studies of case management began to appear (Anthony, Cohen, Farkas, & Cohen, 1988). The results are contradictory: some studies suggest positive client outcomes (Curry, 1981; Goering, Wasylenki, Farkas, Lancee, & Ballantyne, 1988; Modrcin, Rapp, & Poertner, 1988; Muller, 1981; Rapp & Chamberlain, 1985; Rapp & Wintersteen, 1989), and other studies suggest minimal effects on client outcome (Cutler, Tatum, & Shore, 1987; Franklin, Solovitz, Mason, Clemons, & Miller, 1987). Several studies have demonstrated the effectiveness of case management teams, based on the PACT model, that are responsible not

179

only for coordinating all services for a specific group of clients, but also for providing direct services (Bond, Miller, Krumwied, & Ward, 1988; Bond, Witheridge, Wasmer, Dincin, McRae, Mayes, & Ward, 1989; Borland, McRae, & Lycan, 1989; Brekke & Test, 1987; Field & Yegge, 1982; Test, Knoedler, & Allness, 1985). These studies suggest that assigning total responsibility for specific clients to a case management team can reduce drop-out rates, provide more case management to clients who are most disabled, reduce hospitalization, or increase employment and social activity. However, the specific factors that lead to success are still not identified. Some authors emphasize continuity over time, but others emphasize the credibility of the case managers and the visibility of the program as the factors responsible for its success (Grusky et al., 1987; Test et al., 1985).

Problems plaguing outcome research in general also affect the rigor of case management studies—the lack of standardized outcome measures whose validity has been established (Bachrach, 1982a; Farkas & Anthony, 1987; Anthony & Farkas, 1982; Fiske, 1983) and poorly controlled studies in which generalization is not attended to or evaluated. A major problem is the lack of specificity in describing the intervention (Strube & Hartmann, 1983). With respect to this lack of specificity about the intervention, Cohen and associates (1989) embarked on a 3-year technology development process that resulted in a comprehensive case management training technology designed to teach practitioners a person-oriented approach to service coordination. This case management technology teaches the case management skills needed to perform four major case management activities: connecting with clients, planning for services, linking clients to services, and advocating service improvements. Extensive pilot testing of the training technology indicated that case management skills can be taught and measured and their application monitored. (Cohen et al., 1989).

Of particular relevance to the psychiatric rehabilitation approach is a controlled study of case management that used a clearly described intervention based on the rehabilitation case management model (Goering, Wasylenki, et al., 1988). The study reported improved instrumental role functioning, independent living status, occupational status, satisfaction with services, and social functioning among clients receiving the intervention when compared to a

180

matched control group. Another recent study (Rapp & Wintersteen, 1989) reported positive results on goal achievement from use of the personal strengths model. The study evaluated the results of replication of the clearly articulated intervention in a number of sites. Fisher, Landis, and Clark (1988) studied a program that might be described as using the broker model and found that neither linkage/referral nor advocacy corresponded to a decrease in the number of client problems, nor was there an impact on a global measure of client change.

Few studies compare case management with other interventions. Franklin and his colleagues (1987) evaluated a generalist approach to case management compared to the usual aftercare services of a community mental health center (i.e., treatment, referrals, medication). In a pretest/posttest design, more than 400 clients were randomly assigned to experimental and control groups. The findings indicate that for the 138 experimental subjects and 126 control subjects interviewed, after 12 months of participation, those in the case management program received more services, cost more to maintain in the community, and had higher recidivism rates, without any increase in quality of life measures. Bond, Miller, Krumwied, and Ward (1988) also reported a comparative study showing no differences between control groups receiving aftercare services and experimental groups receiving case management services based on the PACT model on such measures as quality of life, medication compliance, involvement in community mental health center programs, and contact with the legal system. However, the PACT clients averaged fewer rehospitalization days and some cost savings.

Outcome measures used in studies of case management and other community support programs remain numerous, diverse, and largely nonstandardized (Bachrach 1982a). Outcome measures include psychiatric symptomatology; psychosocial functioning as measured by residential stability, social autonomy, role performance, employment, and social functioning; and patient satisfaction. Goering, Farkas, et al., (1988) used measures of instrumental role functioning, independent housing status, occupational status, and social isolation. Rapp and Wintersteen (1989) assessed the number of client goals achieved. Bond and associates (1988), Franklin and associates (1987), Bigelow and Young (1983), Field

181

and Yegge (1982), and others have used measures of recidivism and quality of life.

When gathering outcome data, it is necessary to specify the period in which the outcome is expected to occur. Carpenter (1979) suggests that time periods for collecting outcome data should include points preceding the intervention, the beginning of the intervention, the conclusion of the intervention, and at some follow-up date after the intervention is concluded. Both Franklin and associates (1987) and Goering, Wasylenki, et al., (1988) suspect that the usual follow-up period in case management studies (6 months to 1 year) may not be long enough to measure the impact of case management. They also stipulate that, although the linking function of case management is crucial, the number and types of community resources to which a case manager can link clients influences client outcomes.

Incorporating the Psychiatric Rehabilitation Approach into the Mental Health System

System developments in the 1980s, such as the CSS and the emphasis on case management services, should help pave the way for further psychiatric rehabilitation developments in the 1990s. As mentioned in chapter 3, the CSS concept and the psychiatric rehabilitation approach focus on the same target population and are compatible in terms of their respective philosophies. The intervention strategies of psychiatric rehabilitation help to implement the CSS philosophy. The CSS concept of a service system defines the context in which the psychiatric rehabilitation approach should be implemented. However, the CSS concept has not provided a blueprint for how to integrate the psychiatric rehabilitation approach into mental health systems.

A mental health authority that decides to integrate psychiatric rehabilitation into its service system must define its overall philosophy and policies for serving persons with severe psychiatric disabilities in a way that supports psychiatric rehabilitation. In addition, the authority must perform its administrative functions guided by its psychiatric rehabilitation philosophy and policies.

The foundation of the service system is the articulation

of its philosophy and policies. Departments of mental health can begin to integrate a psychiatric rehabilitation approach into their service systems by articulating a philosophy supportive of psychiatric rehabilitation that states beliefs about ideal rehabilitation outcomes. The statements of mission and desired outcomes emphasize helping persons with long-term psychiatric disabilities to be satisfied and to function successfully with the greatest possible independence in the residential, educational, vocational, and social environments of their choice. Rehabilitation values such as independence, competency, freedom of choice, right to support, right to personal satisfaction, normalization, empowerment, individualization, and accountability guide the statement of philosophy.

Policies translate philosophy into guidelines for practice. Policies are formally stated in legislation, regulations, rules, and procedures, and implemented through service delivery. Policies that support rehabilitation state the desired course of action that influence rehabilitation practice (Erlanger & Roth, 1985). Policies on high-priority client populations (e.g., people with severe disabilities), entitlements (e.g., decent housing), ongoing supports (e.g., for as long as needed), high-priority programming (e.g., vocational rehabilitation), required program activities (e.g., skill teaching), and mandatory record keeping (e.g., required functional assessments) are important to support rehabilitations (Cohen, 1989).

A mental health system is administered by a mental health authority (e.g., state department of mental health) that has administrative functions or responsibilities that support service delivery (Barton & Barton, 1983). A mental health authority that integrates rehabilitation into its service system performs its administrative functions in support of psychiatric rehabilitation. Cohen (1989) identified and defined eight administrative functions. These functions include planning, funding, management, program development, human resource development, coordination, evaluation, and advocacy. Table 10–1 lists these eight administrative functions. Each function can be performed consistent with the philosophy and policies that support psychiatric rehabilitation.

The *planning* function of a mental health authority involves designing the service system. Assessing clients and designing new and/or enhanced services are parts of the planning function. A system plan that supports psychiatric rehabilitation identifies the

183

TABLE 10–1   *Eight Administrative Functions of Mental Health Authorities*

- Planning
- Funding
- Management
- Program Development
- Human Resource Development
- Coordination
- Evaluation
- Advocacy

Reprinted from: Cohen, M. R. (1989). Integrating psychiatric rehabilitation into mental health systems. In M. D. Farkas and W. A. Anthony (Eds.), *Psychiatric rehabilitation programs: Putting theory into practice* (pp. 162–191). Baltimore: Johns Hopkins University Press.

overall goals that persons in its target population would most like to achieve (e.g., attend school, work competitively, live independently), assesses client competencies and preferences, and develops or enhances services that develop client skills and supports.

The *funding* function of a mental health authority involves obtaining and dispensing dollars to support services. A mental health authority that supports psychiatric rehabilitation obtains money to fund rehabilitation services (e.g., obtains flexible funding, reallocates custodial funds to rehabilitation services, and/or markets rehabilitation to legislators and other funding sources) and dispenses money to services in a way that allows rehabilitation to happen (e.g., adequate per client expenditures).

The *management* function of a mental health authority involves supervising the operation of services (White, 1981). Contracting, monitoring, and quality assurance are parts of the management function. A mental health authority that supports psychiatric rehabilitation writes and monitors contracts based on rehabilitation indicators such as evidence that every client has chosen an overall rehabilitation goal and is involved in rehabilitation diagnosis, rehabilitation planning, and rehabilitation interventions (Miller & Wilson, 1981).

The *program development* function of a mental health authority involves providing consultation to program administrators. A mental health authority that supports psychiatric rehabilitation provides consultation on designing rehabilitation environments, structuring programs around the rehabilitation process, and developing operating guidelines compatible with psychiatric rehabilitation philosophy.

The *human resource development* function of a mental health authority involves selection and training of personnel. A mental health authority that supports psychiatric rehabilitation makes hiring decisions for rehabilitation services based on the presence or absence of rehabilitation attitudes, knowledge, and skills; writes job descriptions based on the tasks required to rehabilitate clients; and funds both inservice and preservice training programs that develop rehabilitation attitudes, knowledge, and skills in its trainees (Field, Allness, & Knoedler, 1980; Jeger & McClure, 1980).

The *coordination* function of a mental health authority involves assuring interagency collaboration. Development of interagency agreements, joint training opportunities among agencies, and guidelines for linkage between agencies are parts of a system's coordination function. The coordination function facilitates cooperation among different services components within the service system and between itself and the other systems that serve its client population. For example, a mental health system that supports psychiatric rehabilitation works together with the vocational rehabilitation system to support persons with psychiatric disabilities in pursuing their vocational goals (Cohen, 1981).

The *evaluation* function of a mental health authority involves analysis of management information and researching client outcomes. Determining what data to collect, the data collection procedures, and drawing conclusions from the data are parts of a system's evaluation function (Schulberg, 1981). A system that supports psychiatric rehabilitation evaluates the achievement of client rehabilitation goals, the improvement of client competencies (e.g., strengths and deficits in client skills), the increase in environmental resources, and implementation of the rehabilitation process. The data collection incorporates a variety of perspectives (e.g., the perspective of the clients, personnel, and family members). The conclusions drawn from the data should connect client variables, program variables, and outcomes.

The *advocacy* function of a mental health authority involves protecting the rights of the client population (Willets, 1980). Negotiating for more favorable eligibility criteria for entitlements and promoting community acceptance are examples of a system's advocacy function. A system that supports psychiatric rehabilitation advocates for normalized treatment; for residential, educational, vocational, and social opportunities; and for the right of persons with long-term psychiatric problems to choose living, learning, and working environments.

Environmental Context of the Mental Health System

The mental health system functions within an environmental context that includes other delivery systems as well as political and economic factors (Cohen, 1989; Scott, 1985). Figure 10–2 shows how the mental health system is built around the client population and rehabilitation personnel and programs, is supported by mental health authorities, interacts with other services systems, and is influenced by its environmental context. The environmental context either supports or creates barriers to psychiatric rehabilitation. In the United States, the environmental context includes national values, such as democracy (in contrast to authoritarian policy), and states' rights that encourage the delivery system to be individualized by each state and locality according to the needs and wishes of its residents. The political climate and health of the economy influence the availability of funding to support the service system. At the state and federal levels, politics and the economy influence the amount of financial resources and how the resources are allocated. Public opinion greatly influences the environmental context in which a service delivery system functions.

In the United States, the environmental context often provides mixed support to psychiatric rehabilitation. Economic problems, such as a large federal deficit, encourage careful spending of money to rehabilitate persons with disabilities. Public opinion about the visible suffering of persons with severe psychiatric disabilities and homelessness can create pressure on politicians and ultimately on the mental health system either to institutionalize homeless persons with mental illness regardless of their individual needs and goals, or to develop better community services. Some-

FIGURE 10–2 *Cross-Sectional View of
the Mental Health System*

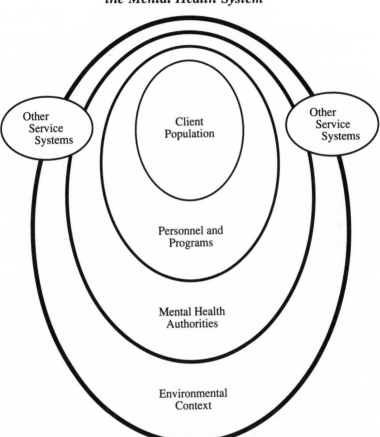

Reprinted from: Cohen, M. R. (1989). Integrating psychiatric rehabilitation into mental health systems.
In M. D. Farkas and W. A. Anthony (Eds.), *Psychiatric rehabilitation programs: Putting
theory into practice* (pp. 162–191). Baltimore: Johns Hopkins University Press.

times the mental health system is forced to prove its effectiveness
and cost benefit. Psychiatric rehabilitation should be able to demonstrate its worth based on cost effectiveness as well as decency to
the client population. For example, vocational productivity and
normalized housing are two rehabilitation goals that meet the values
of both cost effectiveness and decency.

The environmental context of the 1990s began with a legislative initiative that stimulated the continued expansion of psychiatric rehabilitation. Public Law 99–660 provided funds for the development of state mental health plans to establish organized, comprehensive community-based systems of care for persons who are severely mentally ill (NIMH, 1987). This legislation mandated the participation and advice of consumers and their families in the planning process. A state's failure to develop a plan incurred a penalty of 10% of block grant funds (Kennedy, 1989). The legislation specifically mentioned that the state plan describe the rehabilitation services that would be provided. Furthermore, the model state plan developed by NIMH as a technical assistance document is imbued with the philosophy and principles of psychiatric rehabilitation (NIMH, 1987).

Stimulated in part by PL 99–660, in part by the gradual acceptance of the CSS conceptual model, and in part by the relentless advocacy of persons with psychiatric disabilities and their family members, state mental health system planning for how to better serve persons with long-term psychiatric disabilities is increasing. Strategic planning is a key word used by mental health system administrators. Regionalization versus centralization, alternative case management models, changes in eligibility requirements, new funding strategies, needs assessment, and better information systems are the trends in service system planning.

The planning of new and/or enhanced service systems has, of course, occurred before. Most noteworthy has been the development of state hospitals as asylums for mentally ill persons, followed by the deinstitutionalization of state hospitals and the growth of community mental health services. Over the years the tendency in mental health systems planning has been to focus first on such questions as where services should be located, who should administer and fund the services, and how services should be coordinated.

Although these important questions have to be answered, they do not include the first, most important question: *What do clients want and how can they best be helped to get what they want?* Experiential data confirm that persons with severe psychiatric disabilities are similar to all of us in what they want. In hundreds of audiotape interviews of practitioners and clients, the clients repeatedly talk about such goals as wanting to work in satisfying jobs, to live in improved housing, and to develop friendships

(Rogers, Cohen, Danley, Hutchinson, & Anthony, 1986). Unfortunately, neither the hospital-based system nor the community-based system has been very successful in helping persons with severe psychiatric disabilities get what they want (Anthony, Cohen, & Vitalo, 1978; Bachrach, 1983; Talbott, 1983).

Past Pitfalls in Mental Health Planning

The current round of mental health systems planning is an opportunity to design a more responsive service system, but it is also riddled with pitfalls from the past. The following six pitfalls could get in the way of planning a system that helps clients get what they want (Cohen & Anthony, 1988).

1. *Lack of values.* Systems planning has been almost devoid of values. Occasionally people talk about values being important, but rarely specify what the values are. Yet explicitly stated or not, values underlie all our service system planning. Perhaps the hesitancy to state the values underlying the planning of mental health service stems from the desire to avoid controversy. Certainly mental health systems planning benefits from consensus support, but at the point of designing services, the underlying values (whatever their source) need to be stated. For example, three values that can guide service planning are maximizing choice, increasing competency, and providing support (Anthony, Cohen, & Cohen, 1984). Whether these or other values are selected as the foundation of systems planning, a system is as value-able as its values (Cohen & Anthony, 1988).

2. *Lack of focus on client goals.* Mental health systems planning has traditionally focused on achieving goals that change the service system. Community-based services and service utilization are familiar system goals. However, even though they are important objectives for improving the delivery of services, they cannot drive systems planning. Systems planning must be driven by goals that come directly from the client population (e.g., satisfying jobs).

Common objections to focusing on client goals are that clients are too unmotivated, unrealistic, and ill to select constructive goals or that the existing resources are too limited to allow client choice. However, by purposefully listening to clients discuss their

189

goals, planners will learn that most persons with psychiatric disabilities can and do express constructive goals. Another objection to focusing on client goals is the pressure to respond to the goals of such other populations as professionals, neighborhood groups, unions, and legislators. Although the pressure to respond to the goals of other populations makes it more difficult for the system planners to focus on client goals, they must have the courage to do so (Cohen & Anthony, 1988).

3. *Lack of focus on clients' perceived needs*. Systems planning has not been based on clients' perceived needs. Typically, when assessing needs, service providers usually assess the need for predetermined services such as the need for day treatment services or they assess service needs by assessing clients' levels of functioning. In a client-driven system, on the other hand, clients assess their perceived needs for specific assistance to alleviate the distress they experience, for example, a need for assistance in finding independent housing with compatible people (Cohen & Anthony, 1988). Service systems can only be functionally designed if the clients' perceived needs for assistance are known and used in systems planning.

4. *Lack of focus on preferred level of intervention*. Some clients want and use a variety of services (e.g., medication, psychotherapy, psychosocial rehabilitation), and other clients want little or no services. In planning service systems, we have not focused on the client's preferred level of intervention, how much help they really want. Service systems can offer varying amounts of support to meet the preferences of its consumers, rather than requiring them to take all or nothing (Cohen & Anthony, 1988).

5. *Lack of substance*. Systems planning has focused on the configurations of services without paying close attention to the substance of services. A particular configuration of services is seen as a blueprint for the design of a better service system. However, the essence of service delivery—what intervention is provided by whom in which environment and for what purpose—has been missing (Cohen & Anthony, 1988). As discussed in chapters 8 and 9, it is possible to describe the mission, the program structure, and staff competencies required for improved service delivery—the substance of a service system (Cohen, 1989; Farkas, Cohen, & Nemec, 1988).

6. *Lack of hope.* Thus far, our vision has been limited in systems planning. Planning has concerned primarily the issue of location. The mission has been to move patients from the hospital to the community and then keep them there. Previous systems planning has not demonstrated a belief in the rehabilitation potential of persons with long-term psychiatric disabilities. Current systems planning can incorporate hope in its planning of services by designing rehabilitation services that give clients the opportunity to develop life-enhancing skills and supports. The vision need not be simply of community care or community maintenance, but rather of client involvement and growth in the community of choice.

In summary, mental health service systems are at a crossroads. The states are planning improved services in the community. The hospital is now seen as a part of this community system. There is the opportunity to plan not only a community-based service system, but also a system responsive to the goals of its clients. States can build service systems that avoid the pitfalls of the past if systems planning begins with an understanding of what the clients of the system want.

Unless service system plans for mental health articulate their underlying values, are driven by client goals, begin with an assessment of client needs and preferences, detail the program substance, and are based on a hopeful vision of what is possible, the new mental health systems may be no better than previous ones (Cohen & Anthony, 1988). Clients and their families will be asking, What is the benefit for the users of these new services? Planners will need a response to this fundamental planning question. The planning pitfalls of the past must this time be avoided.

Concluding Comment

Regardless of how well crafted the plan is, how steeped it is in the appropriate philosophy, how up-to-date the planner's knowledge is, or how strategic the planning process is, the success of the plan is determined by what it does for the persons being served. A plan can do nothing unless the practitioners and administrators who implement the plan are expert in the technology needed for successful implementation (Anthony, Cohen, & Kennard, 1989).

As pointed out in this chapter, most current service-system planning focuses on the *process* of planning rather than on the *substance* of the plan. A strategic planning process has been adopted by many states (Goodrick, 1988). The next essential step is to change the substance of planned services (Cohen, 1989). Technological advances in the field have resulted in knowledge that can be applied to developing the program structures and staff competencies required for improved services to clients (Anthony, Cohen, & Farkas, 1987).

The means of helping persons with severe mental illness are no longer a mystery. Developing technology with its applied knowledge can be used easily in a replicable fashion to change program structures and staff competencies in ways that will decrease clients' symptoms and improve clients' skills, supports, and role performance (Anthony, Cohen, & Kennard, 1989).

Technologies vary considerably with respect to their level of detail, but the more fully developed technologies have been developed to such an extent that instructors can teach them, service providers can use them, administrators can monitor them, researchers can evaluate them, consultants can disseminate them, and/or consumers and family members can observe them. Service models, training curricula, and consultation strategies exist. For example, these technologies range from how to create a psychosocial rehabilitation center, how to develop consumer operated self-help programs, how to teach medication management skills, how to set goals, how to conduct functional assessments, how to teach skills, how to connect with clients, and how to perform case management activities (Anthony, Cohen, & Kennard, 1989).

The adoption and use of technology are to a certain extent dependent on how the technology is described and packaged (Muthard, 1980). Some of the newer technology useful in improving services to persons with long-term psychiatric disabilities is defined in great detail and packaged to ensure easy use. In other instances the use of some of the new technology depends on the participation of the developers in the technology transfer process. Also important to technology transfer are the characteristics of the users of the technology (Gomory, 1983). The more knowledgeable and skilled the user of the technology, the less well developed the technology needs to be because the user is capable of adding the necessary details.

Attempts to implement systems planning based on a rehabilitation philosophy will either lead to the adoption of new technology or the discarding of the rehabilitation philosophy (Anthony, Cohen, & Kennard, 1989). If new technologies are adopted, not only will the planning of services change, but also, more important, the delivery of services.

11

Technology for Change

> *Much of the change we think we see in life is merely truth moving in and out of favor.*
>
> *Robert Frost*

*I*f psychiatric rehabilitation is to remain in favor, then improvements must be made in how personnel are trained, programs are developed, and service systems are designed. The practice of psychiatric rehabilitation requires unique practitioner skills, knowledge, and attitudes; innovative programs; and a service system that supports psychiatric rehabilitation. A training and consultation technology currently is being developed (Center for Psychiatric Rehabilitation, 1989) to equip agencies with skilled rehabilitation practitioners and substantive programs.

Parts of this chapter were excerpted with permission from:

Anthony, W. A., Cohen, M. R., & Farkas, M. (1987). Training and technical assistance in psychiatric rehabilitation. In A. Meyerson & T. Fine (Eds.), *Psychiatric disability: Clinical, legal, and administrative dimensions.* Washington, DC: American Psychiatric Press.

Cohen, M. R. (1989). Integrating psychiatric rehabilitation into mental health systems. In M. D. Farkas & W. A. Anthony (Eds.), *Psychiatric rehabilitation programs: Putting theory into practice* (pp. 162–191). Baltimore: Johns Hopkins University Press.

This technology enables the field to change its personnel, its programs, and its service systems most effectively.

The Experience of Change

Becoming an expert in the practice of psychiatric rehabilitation *cannot* be accomplished overnight. Expertise is *not* automatically acquired when a person is hired by a rehabilitation program. Developing expertise in the practice of psychiatric rehabilitation takes a great deal of time and hard work. A trainee describes her experience in psychiatric rehabilitation training thus:

> My extensive psychosocial background led me to believe that I was as psychosocial as one would get. Then came Psychiatric Rehabilitation Skills Training! Doing it, however, involves a much harder task than the initial material indicates. Quickly one learns that "simple" does not necessarily mean "easy."
>
> There were three grueling weeks of training, audiotapes of sessions of clients and me to be critiqued by our trainers, teaching sessions with my staff on material I was still struggling to learn (52 hours of teaching since April), study groups, critiquing my staff's tapes, etc., etc.! The results from all this I am finding, however, to be tremendous.
>
> Psychiatric Rehabilitation offers great value to our clients and helps us put into action what conceptually we all know, believe, and even think we are doing: in actuality we have just scratched the surface. The model forces us to carry our psychosocial philosophy to its fullest extent . . . .
>
> Implementation of this model is very challenging. It involves extensive training for staff and program development for application of the model in individualized programs. The single most important factor, however, is administrative support. Implementation for me would have been virtually impossible without agency commitment to the long-term benefits . . . .

The next step should be to provide a consistent philosophical and programmatic approach for the client within the same service system. Sounds like a good idea to me, but a lot of work! We can do it! (Hillhouse-Jones, 1984, pp. 1, 8).

As discussed in chapter 8, most practitioners have not been formally educated or trained in psychiatric rehabilitation knowledge and skills (Friday, 1987). Instead, they learned about rehabilitation through their firsthand experiences in psychosocial and rehabilitation settings and from the leaders in the field, such as Beard, Grob, Dincin, Rutman, and others. However, as the popularity of the psychiatric rehabilitation concept increased and more people became involved in its practice, the need for technical assistance (i.e., personnel training, program consultation, and systems consultation) became more apparent. In 1977, Anthony facetiously characterized the present state of psychiatric rehabilitation as "the development of nontraditional psychiatric settings for the purpose of using traditional psychiatric techniques by traditionally trained personnel" (Anthony, 1977, p. 660). Clearly, psychiatric rehabilitation training and consultation were needed.

The need for psychiatric rehabilitation training and consultation has been preceded by the acceptance of the concept of psychiatric rehabilitation and the creation of rehabilitation settings. Consensus on the philosophy and principles underlying the field is emerging. In addition, several research studies have shown that the two major interventions of psychiatric rehabilitation—teaching skills to clients and developing environmental supports for clients— are related to various types of client rehabilitation outcome (e.g., more frequent and higher levels of client independent living and vocational functioning). In essence, the articulation of the philosophy of psychiatric rehabilitation, the empirical studies of rehabilitation, and the nationwide development of psychosocial rehabilitation settings have resulted in further definition of the psychiatric rehabilitation approach. The articulation of the psychiatric rehabilitation approach and the identification of those practitioner skills and program elements required for its implementation have made possible the provision of technical assistance designed to change practitioners, programs, and service systems.

A Training Technology to Train Practitioners

The essential skills of a psychiatric rehabilitation practitioner can now be observed and measured. Consequently, a training technology to teach practitioners these skills is being developed. As noted in chapter 8, most current practitioners of psychiatric rehabilitation were not trained in psychiatric rehabilitation skills as students, and most universities have still not adopted this role. Thus, practitioners often acquire psychiatric rehabilitation training after they become employed in a setting that serves persons with severe psychiatric disabilities.

Certain skills are useful for the practitioner to possess no matter what the particular program setting (e.g., clubhouse, community mental health center, inpatient unit, state vocational rehabilitation office). Anthony, Cohen, and Pierce (1980) co-authored six books for the purpose of teaching practitioners some of the previously identified skills of psychiatric rehabilitation. The training program was evaluated in a 150-hour pilot training program for student interns studying rehabilitation counseling and practicing mental health and vocational rehabilitation practitioners. The training was built around the skills presented in six practitioner books. The evaluation of the training indicated that the skills of psychiatric rehabilitation can be successfully learned and measured and that trainees considered these skills to be important in the performance of their jobs. Furthermore, clients of practitioners who scored high on written assessments of these skills were more apt to report feeling involved, understood, and taught new skills (National Institute of Handicapped Research, 1980). The Center for Psychiatric Rehabilitation has since evaluated the training of hundreds of practitioners in psychiatric rehabilitation skills (Center for Psychiatric Rehabilitation, 1984; Goering, Wasylenki, et al., 1988; Rogers et al., 1986) and has incorporated skills training into masters and doctoral degree programs at Boston University (Farkas, O'Brien, & Nemec, 1988).

Development of a Psychiatric Rehabilitation Training Dissemination Strategy

Although the necessity and value of training had been documented, there was no mechanism in place for large-scale dissemination of the training to mental health practitioners. A training dissemination strategy was clearly needed.

Reviews of the research concerning dissemination and use of new technology provided a starting point for developing a workable strategy (e.g., Caplan, 1980; Fairweather, 1971; Glaser & Taylor, 1969; Hamilton & Muthard, 1975; Havelock, 1971; Havelock & Benne, 1969; Muthard, 1980; Pelz & Munson, 1980; Soloff, 1972; Switzer, 1965). The literature suggests that training is ideally an ongoing process built into an organization's operation. It is preferably viewed by all those affected as crucial to the maintenance and growth of the organization. The active involvement from the beginning of key personnel, especially administrators and practitioners, is vital (Muthard, 1980). Such initial involvement creates a sense of ownership, identification, and purpose. This situation is enhanced when the persons involved feel the need for the training (Glaser & Taylor, 1969; Pelz & Munson, 1980).

Based on these research findings as well as on training experience, a training-of-trainers strategy has been developed to disseminate psychiatric rehabilitation skills to practitioners. The strategy is special in that it incorporates three key ingredients based on the literature review:

1. Careful selection of service settings in which to disseminate the training, and careful selection of the personnel who are to be trainers.
2. Training of trainers in both the clinical practice of psychiatric rehabilitation skills and the teaching of these skills to others.
3. On-site assistance to trainers in planning and implementing the training in their service settings.

In an NIMH-funded evaluation of the training dissemination strategy with mental health agencies, 100 agencies from around the country volunteered to be training sites, 9 of which were selected. Representatives from these agencies were trained in the psychiatric rehabilitation skills and in how to teach these skills to others. These new trainers then trained their agency staffs in psychiatric rehabilitation skills, and the agency staffs used these skills with their clients. Data analysis indicated the viability of the strategy for facilitating practitioner training and use of the psychiatric rehabilitation skills (Rogers et al., 1986). Since the development and research of the training-of-trainers strategy, Center for Psychiatric Rehabilitation staff have trained in some part of psychiatric rehabilitation technology) more than 100 trainers representing more than 50 service settings including state departments of mental health, community mental health centers, state psychiatric hospitals, psychosocial rehabilitation centers, and private hospitals.

Training Technology that Supports the Training Dissemination Strategy

The training of trainers, as well as the training of practitioners, requires special expertise and technology. The technology that supports the training dissemination strategy includes practitioner books, trainer packages, and master trainer guides. The various practitioner books (Anthony, Cohen, & Pierce, 1980; Cohen, Danley, & Nemec, 1985; Cohen, Farkas, & Cohen, 1986; Cohen, Farkas, Cohen, & Unger, 1990) introduce the practitioner to the skills of psychiatric rehabilitation. After training, practitioners use the books to review what they learned in their training. The trainer packages (M. R. Cohen et al., 1985, 1986, 1990) contain detailed lesson plans and training aids (videotapes, audiotapes, transparencies) for use during training sessions. The packages are important because they save trainer preparation time and provide a tested program with attractive training aids. Some guides provide master trainers (i.e., the persons who train the trainers) with information about how to develop new trainers.

In summary, the training technology rests on the firm foundation of the psychiatric rehabilitation approach. Training programs teach the essential psychiatric rehabilitation skills to psychiatric rehabilitation practitioners. The dissemination of training programs involves a training-of-trainers strategy that prepares trainers and master trainers to use training technology to teach practitioners and trainers.

A Consultation Technology to Change Programs

The implementation of psychiatric rehabilitation depends on the practitioners' ability to use their competencies effectively on a regular basis. A rehabilitation program provides the organizational support to practitioners in diagnosing, planning, and intervening with their clients.

The essential characteristics of a program based on a psychiatric rehabilitation approach have been identified (see chapter 9). However, even though more is known about these characteristics, creating a new psychiatric rehabilitation program or changing a traditional program into a rehabilitation program is very difficult (Carling & Broskowski, 1986). Many rehabilitation-oriented programs still do not use the new knowledge (Farkas, Cohen, & Nemec, 1988). The development of a psychiatric rehabilitation program usually requires changes in the physical environment, funding, daily operation, staffing, mission, program structure, and record keeping, as well as the creation of new rehabilitation environments (Marlowe, Spector, & Bedell, 1983). Often, staff assigned the responsibility of program development lack the knowledge and skills to make these changes.

In the past, providing technical assistance has helped agencies to develop or change. Technical assistance involves the transfer of specific knowledge and skills (Domergue, 1968; Havari, 1974; Sufrin, 1966) from a donor (i.e., technical expert) to a recipient (i.e., technical nonexpert). Nemec (1983) provides a discussion of the distinction between technical and other forms of assistance (e.g., financial assistance, technical cooperation). In essence, technical assistance involves both consultation and training.

Consultation is defined as the transfer of technical knowledge concerning psychiatric rehabilitation, program development, and systems planning. Training is defined as the teaching of psychiatric rehabilitation attitudes, knowledge, and/or skills for the purpose of personnel development. The format for program consultation is usually a one-to-one, on-site relationship, whereas training is often delivered in off-site group meetings. There are endless examples of on-site consultation in the mental health field. For example, NIMH frequently provides consultation for new program developments (e.g., development of community mental health centers and community support programs).

Two formal providers of technical assistance are the Fairweather Lodge (Fairweather, 1980; Fergus, 1980) and Fountain House (Propst, 1985). Fairweather Lodge studied the effect of their program consultation and suggested some recommendations relevant to providing technical assistance. These recommendations included the importance of involving a number of program staff and power blocks in a discussion of program change and forming an action group to identify the concrete action steps that need to be taken (Fairweather, 1980).

Since 1977, Fountain House has conducted national training and program consultation to develop new clubhouses and to strengthen existing clubhouses. Many of the participating programs have been day treatment programs of community mental health centers. After receiving a pretraining site visit from Fountain House staff, program staff receive 3 weeks of training at Fountain House, essentially serving as apprentices. After the training, consultation is provided on- and off-site for as long as needed (Propst, 1985). As a program trainer stated, "colleagues in training become able to replicate, individualize, and improve on a service model which enhances the quality of its members' lives" (Shoultz, 1985, p. 2).

Making a Program More Compatible with a Psychiatric Rehabilitation Approach

A psychiatric rehabilitation program consultant needs specific knowledge (i.e., the elements of a psychiatric rehabilitation program) and unique consultant skills (i.e., how to create a psychiatric rehabilitation program). The Center for Psychiatric Rehabilitation has conceptualized this consultation process as four activities: (1) determining program readiness, (2) assessing the program, (3) proposing program change, and (4) creating the program change (Center for Psychiatric Rehabilitation, 1989). The process emphasizes a careful diagnosis of the program (e.g., its mission, structure, and network of environments) to evaluate the program's compatibility with the psychiatric rehabilitation approach and its potential for change.

Based on the determination of readiness and the assessment of program strengths and deficiencies, the consultant proposes a plan for change. The plan specifies the consultation interventions and timelines for achieving the goals. Next, the interventions are

202

implemented to create program change. These interventions can include developing a new mission statement, writing new policy statements, designing a new program structure, revising record keeping, modifying job descriptions, and creating new living, learning, social, or working environments.

The ingredients of psychiatric rehabilitation program consultation were studied when the New Jersey Division of Mental Health and Hospitals funded a technical assistance project to help the state's 54 partial-care agencies deal with the multitude of service issues encountered in rehabilitating persons with long-term psychiatric disabilities. Four consultants and a project supervisor were selected because of their experience in psychiatric rehabilitation programs, their technical knowledge, and their consultation skills (Kelner, 1984). The consultants were located at a psychosocial rehabilitation center that was also funded by the state. The project supervisor received technical assistance from the Center for Psychiatric Rehabilitation for developing a program consultation process that included three phases: (1) assessment of partial-care programs, (2) development of an individual technical assistance plan, and (3) provision of technical assistance interventions such as common issue training and on-site consultation. A further description of the New Jersey Partial Care Technical Assistance Project is reported in Borys and Fishbein (1983), Kelner (1984), and Fishbein (1988). Evaluation of this consultation process has indicated positive results. For example, major changes from pre- to posttest were seen in the agencies' documentation of client skill strengths and deficits and in the prescription of specific interventions relevant to client goals. In addition, client involvement was more routinely documented and more assessments were made of the clients' environmental supports.

During the past 5 years, more and more programs have used training and consultation strategies to help them improve the delivery of psychiatric rehabilitation to their client population. A review of the experiences of some of these residential, vocational, educational, and social programs indicates the degree to which long-term intensive dissemination strategies are actually necessary to incorporate innovations into the daily practice of program settings (Farkas & Anthony, 1989). Given the lack of consultants with these specific skills, additional training technology is needed to

203

increase the pool of resources for improving programs (Center for Psychiatric Rehabilitation, 1984).

To meet this need, a training program for teaching the specialized knowledge and skills of psychiatric rehabilitation program consultation has been developed (Center for Psychiatric Rehabilitation, 1989), demonstrated, and evaluated (Nemec et al., 1990). Four community agencies from West Virginia were preassessed as to their compatibility with a psychiatric rehabilitation approach, using the program assessment procedures described in chapter 9. A two-part training strategy to train program consultants was then implemented. Initially, 10 people were trained in practitioner skills and then used these skills with actual clients. At the completion of this practitioner training, 4 of the 10 went on to the second phase of program consultation training, which focused on learning program consultation skills. This 18-day training program covered such topics as how to conduct program assessments, how to plan for program change, and how to create psychiatric rehabilitation programs within an existing agency's structure.

After the consultants had been trained, the four agencies received different kinds of consultation. Two agencies received formal consultation (8 days of consultation spread over a 10-month period), one agency received informal consultation (5 days over a 10-month period), and one agency received only feedback on the preassessment. The consultations involved the newly trained consultants working with agency leaders, who in turn worked with implementation committees. The consultations focused on the agencies' missions and how the agencies were structured to deliver the psychiatric rehabilitation process of diagnosing, planning, and intervening. Following the consultation period, program assessment was repeated on the three sites that received the formal and informal consultation. In addition, the amount of perceived change was rated by the consultants, the agencies, and the state mental health authority (see Table 11–1).

The posttest ratings, like the pretest ratings, were made by trained evaluators from the Center for Psychiatric Rehabilitation. The program assessment instruments indicated positive change in all three agencies receiving consultation. Program missions

TABLE 11–1 *Agency Change Ratings*

| | Assessed Change | Perceived Change | | |
	Instruments	Consultant	Agency	State
Agency 1				
Mission	o	+ +	+	o
Diagnosing	+	o	+	o
Planning	o	+	+	+ +
Intervening	+	+	+	+
Agency 2				
Mission	o	+ +	+	+
Diagnosing	+	+ +	+	+
Planning	o	+ +	o	+
Intervening	o	+ +	o	+
Agency 3				
Mission	o	+	+	+
Diagnosing	+	+	+ +	+
Planning	+ +	+	+ +	+ +
Intervening	o	+	+ +	+ +

Explanation of rating: + + = much or great change
+ = some change
o = little or no change

did not change, but program structure did change in each agency, with diagnosis showing a consistent improvement. Ratings of perceived change ranged from little or no change to great change depending on the source of the ratings. Rating the consultation procedures as most successful were the consultants themselves, followed by the agency administrators, and state mental health authority.

The variation in perception of change, as well as the difference between perceived change and assessed change, indicate the importance of supplementing measures of perceived change with more objective assessment measures. This supports the results of an earlier study of 40 agencies that demonstrated discrepancies between assessed congruence with psychiatric rehabilitation and the beliefs of program administrators about the rehabilitation orientation of their programs (Farkas, Cohen, & Nemec, 1988). In that study, the program administrators stated that skills were assessed as part of the agencies' assessments of clients, yet only 30% of the programs actually evaluated client behaviors that could be labeled as skills. Additionally, 61% of the program administrators reported conducting structured, skill training sessions, when in actual practice they conducted discussion groups and group therapy meetings.

The results of the study of the training of program consultants in West Virginia are similar to the earlier study reported by Fishbein (1988), who found the greatest change in rehabilitation diagnostic procedures. That study showed that rehabilitation planning changed least but was the strongest area of the programs at the initial assessment. The West Virginia agencies were initially strong in rehabilitation planning as well, as were the 40 agencies assessed in the study mentioned earlier (Farkas, Cohen, & Nemec, 1988). Similarly, in the initial assessment, all three studies showed rehabilitation intervention to be the weakest element in the agencies' programs.

Planning may have been consistently stronger for the assessed programs for several reasons. State funding and insurance reimbursement regulations require clear objectives and ongoing review of plans and often insist on individualized plans that indicate client agreement. These characteristics are all desirable in a psychiatric rehabilitation program (Farkas, Cohen, & Nemec, 1988).

In contrast to planning, rehabilitation diagnosis and interventions are most influenced by a traditional psychotherapeutic orientation. Diagnosis often draws from *DSM III* or *DSM III-R* (American Psychiatric Association, 1980, 1987), a diagnostic system largely irrelevant to psychiatric rehabilitation (Anthony & Nemec, 1984). Interventions based on the clinical therapeutic techniques still taught in most professional mental health disciplines might be described as round holes into which persons with severe psychiatric disabilities—the square pegs—are expected to fit, with little or no success (Mowbray & Freddolino, 1986). Changes in

206

the techniques used may be difficult to implement due to negative staff attitudes toward working with this population (Mowbray & Freddolino, 1986; Stern & Minkoff, 1979). A Michigan case study (Mowbray & Freddolino, 1986) identified an additional area of difficulty in changing the interventions offered: staff tended to deliver services so as to meet their own professional needs and desires, but not necessarily those of their clients. Finally, interventions may be difficult to alter because they are not monitored closely through the record-keeping process. Practitioners are not often required to document and justify their interventions in detail. In contrast, both diagnosing and planning result in a written product that both guides and monitors the practitioner's performance.

The positive changes assessed and perceived in the West Virginia study demonstrate that even minimal program consultation can have an impact. The consultation time was no more than 8 days with any agency, and the program assessment required another 2 days. The consultation provided during the 8 days included setting consultation goals based on the assessment results. At least 25% of the total consultation time related directly to conducting and discussing the program assessment. Even though it is not possible to separate the impact of the program assessment process and results from other factors promoting change, presenting an agency with this type of assessment information can, in itself, produce change (Nadler, 1977).

Requests for technical assistance should come from the agency, because it believes it needs assistance. Staff attitudes should also reflect this belief. Support from the agency's service system is also critical to the success of technical assistance. Weaknesses in any of these areas can present significant barriers to the success of the consultation.

Along with technical assistance, provision should be made for continued assessment of the effectiveness of the strategies and for identifying and addressing unanticipated barriers. Once the technical assistance is completed, an objective evaluation should be conducted of how well the desired goals and objectives have been achieved. In addition, the subjectively perceived impact of technical assistance should also be assessed (Nemec et al., 1990).

A System Consultation Technology to Change Service Systems

As explained previously, the psychiatric rehabilitation approach must become an integral part of the practice of the personnel, the programs that structure the various environments, and the service system that organizes the services in a specified geographic area. System consultation has a better chance of producing changes in client outcome, because of the advances in practitioner training and program consultation.

System consultation can be more effective because the attitudes, knowledge, and skills of personnel involved in psychiatric rehabilitation have been described and translated into a training technology. System consultation can be more effective because the program mission, structure, and network of environments that support the rehabilitation process can be described and translated into a technology for program consultation. System consultation can thus have a substantive direction. The target population, philosophy, policies, and functions of the mental health authority can be designed so that the personnel and programs in the service system are effectively supported in providing psychiatric rehabilitation.

A mental health authority that integrates psychiatric rehabilitation into its service system defines its target population, philosophy, policies, and functions for serving the psychiatrically disabled in a way that supports psychiatric rehabilitation. The community support system discussed in chapter 10 is a good example of a design compatible with psychiatric rehabilitation (Cohen, 1989). In addition, a burgeoning literature focuses on various elements of system change, such as financing, planning, programming, and staffing (e.g., Carling, Miller, Daniels, & Randolph, 1987; COSMOS, 1988; Harris & Bergman, 1988a; Jerrell & Larsen, 1985; Lehman, 1987; Mosher, 1983; Santiago, 1987; Talbott, Bachrach, & Ross, 1986; Telles & Carling, 1986). Knowing how systems change aids in planning how to incorporate psychiatric rehabilitation into a system. As states move to implement PL 99-660, which required incorporating a rehabilitation philosophy into state plans (NIMH, 1987), there will be even more opportunities to learn about systems change relevant to psychiatric rehabilitation (Anthony, Cohen, & Kennard, 1989).

The system begins its integration of psychiatric rehabilitation with specification of the target population for psychiatric rehabilitation. In the broadest context, the target population for psychiatric rehabilitation are people recovering from psychiatric impairment *who are not functioning at the level they would like to function* (Cohen, 1989). Specific mental health systems may need to differentiate the population they can serve further by describing demographic criteria, such as residential status, age, or income. In addition, in times of funding shortages, systems may need to specify the severity of impairment in their statement of target population (Bachrach, 1980). In the ideal system, however, all people who can and want to benefit from rehabilitation, regardless of the level of their impairment, are the target for rehabilitation.

Process of Systems Change

Pierce and Blanch (1989) have stated six conclusions about the systems-change process based on their experience of trying to get the Vermont mental health system to incorporate psychiatric rehabilitation. They conclude that changing the mental health system to incorporate psychiatric rehabilitation is unpredictable, requires service agencies to have at least one staff member ready and willing to serve as an advocate, requires flexibility in the choice of strategies, requires financial support tied to rehabilitation outcomes, and is slow and variable. Their findings are consistent with the literature on the dissemination and use of new technology (Caplan, 1980; Hamilton & Muthard, 1975; Havelock & Benne, 1969; Soloff, 1972; Zaltman & Duncan, 1977).

Most critical to system change is that the change process be driven by a systemwide mission that is client oriented and rehabilitation focused (i.e., a mission directed toward successful client functioning and satisfaction in environments of the client's choice with increased client independence). The change process for service systems is characterized by the following five key steps according to Cohen (1989):

1. A rehabilitation mission for a clearly defined target population is stated.
2. Policies that translate the mission into guidelines for practice are written.
3. Rules, regulations, and procedures are written consistent with these rehabilitation policies.
4. Rehabilitation outcomes based on the mission and policies become goals that the system works to achieve.
5. All major system functions (e.g., planning, funding, management, human resource development, coordination, evaluation, and advocacy) support achieving the rehabilitation outcomes.

Principles of System Change

As with most change, a significant lag in time between the development of knowledge about rehabilitation services and use of this knowledge within mental health systems is expected. Muthard (1980) and Glaser and Ross (1971) maintain that promoting cognitive awareness alone is insufficient to bring about change in practice. Jung and Spaniol (1981) summarized the findings from a comprehensive review of the dissemination and utilization literature and distilled 15 principles relevant to systems change. These principles focus on the ideal product (e.g., credible, observable, relevant, advantageous, easy to use, and compatible), ideal process (e.g., ongoing assessment, involvement of users, strategy development, and support for users) and the ideal context (e.g., readiness for change, adequate resources, and a felt need within the environment). Based on an analysis of the systems change experience in New Jersey (Fishbein & Cassidy, 1989), Vermont (Pierce & Blanch, 1989), and West Virginia (Nemec et al., 1990), and on the literature on systems change, Cohen (1989) extracted the following 10 principles for changing a mental health service system to integrate psychiatric rehabilitation programming:

1. Changing the system centers on the needs and preferences of the clients to be served.
2. Changing the system is facilitated by an assessment of compatibility and readiness for change (e.g., a felt need for change) within service settings.

3. Changing the system is facilitated by the teaching of new skills to staff and/or by supporting staff use of existing skills within service settings.
4. Changing the system is facilitated by peer models in the service settings.
5. Changing the system is facilitated by creating a supportive environment within service settings including selecting rehabilitation-oriented management and developing compatible program structures.
6. Changing the system is facilitated by supportive functioning of the state departments of mental health, including supportive planning, program development, human resource development, management, coordination, funding, evaluation, and advocacy.
7. Changing the system is facilitated by adopting a rehabilitation technology that is credible, observable, relevant, compatible, understandable, and accessible.
8. Changing the system is facilitated by positive interpersonal relationships between personnel in the state department of mental health and personnel in the service settings characterized by tailoring to individual agency needs and shared responsibility.
9. Changing the system is facilitated by assessment and intervention with other relevant service systems to assure their support of the rehabilitation mission.
10. Changing the system is developmental and takes sufficient time (e.g., usually 3 or more years).

In summary, integrating rehabilitation into mental health systems is a difficult challenge. The process of change is laborious and requires careful definition of the target population, the system's mission for the population, and the policies and procedures that support rehabilitation. All functions of the system need to be redesigned to support psychiatric rehabilitation.

Concluding Comment

In summary, there is burgeoning knowledge about psychiatric rehabilitation. No longer must personnel learn psychiatric rehabilitation practice by trial and error—often at the expense of clients. A psychiatric rehabilitation technology, in which practitioners can become expert and that guides the development of programs and systems, is being developed. The training and program consultation technologies needed to disseminate the psychiatric rehabilitation technology are being developed. The growth of knowledge about changing service systems, along with the technology to train practitioners and develop programs, will give power to the psychiatric rehabilitation field's good intentions and substance to its hopes.

12

Vision
of the Future

*Happy are those who dream dreams, and are willing
to pay the price to make them come true.*

Suenens

Researchers and practitioners in the field of psychiatric rehabilitation are sometimes reluctant to discuss their visions of the future. In contrast, persons who practice and research psychiatric treatment often talk about the future. They speak of possible breakthroughs due to new technology like PET scans and CAT scans, of advances in medication, of prevention, and of cure. They have a vision toward which they are working. They point out the obstacles, the difficulties, the required time and resources—but they do have a vision.

Psychiatric rehabilitation professionals, on the other hand, often talk about how far we have come, what our current resources and opportunities are, and how far we still need to go. But where is the vision, the hopes, the long-term dreams? We may be so worried that we will not produce what we promise that we unduly restrict our vision and stifle our dreams. There is a difference between raising false expectations and putting forth a vision toward which to work. If we continue to work toward and advocate that vision, then the vision is not misleading—it is encouraging. A vision begets not a false promise but a passion for what we are doing.

This last chapter is written more from the heart than from the head. It contains no references. We the authors, like many of the other constituencies in the field (i.e. consumers, families, practitioners, educators, administrators, and researchers), have earned the right to dream. We hope this vision is a shared vision.

Getting Ready to Dream

In order to open up our minds to what could be, we need to put to rest the concerns that truly should be concerns of the past. We cannot look to the future if we are facing in the wrong direction. As Winston Churchill said, "If we open up a quarrel between the past and the present we shall find we have lost the future." We get ready to dream by first putting the past to bed. These six issues must be seen as yesterday's issues so that we can face squarely tomorrow's vision.

Issues of the Past

We need to stop asking whether there is a role for consumers and start asking consumers what roles they want. Fortunately, under the leadership and prodding of NIMH's Community Support Program, more and more mental health and rehabilitation professionals are realizing that consumers do have a role and that they will ultimately determine their own role. Every other disability group has developed self-help and advocacy functions. The professionals must stay in touch with what the persons with the disabilities and their family members believe to be helpful to them. In the future, consumers on policy-making boards and consumers working as practitioners or educators will become less of a novelty. The door has been opened—consumers are in. We must stop fretting, questioning, or directing—and start listening.

We need to refrain from arguing about the merits of hospital versus community care and identify the functions each performs best. The debate about *where* rather than *what* has consumed too much energy already. Consumers of mental health

214

services, just like consumers of other health services, prefer to live in the community. The question is, How can hospital and community services help them to be more satisfied and successful in the community?

In the final analysis, the hospital is a part of the community. It has a role to play in helping persons live in the community, not in keeping them out. Yet the community-based professional sometimes treats the hospital as if it were another planet. Persons who become hospitalized are forgotten by people who are not hospitalized. Often no one visits, telephones, or writes letters to them—as if they had indeed gone to another planet! Community-based practitioners must follow their clients into the hospital, and the hospital must keep its focus on helping persons return more capably to the community. Both hospital and community programs are properly community focused. Both should enable clients to live, learn, socialize, and work where they want.

In the past, community-based professionals assumed their programs were better than hospital programs because they were, after all, in the community. This assumption is not and was never true. Programs in the hospital can also operate from a rehabilitation orientation and, by working collaboratively with community-based programs, have a positive impact on client rehabilitation outcome. No matter *where* the setting, professionals must focus on *what* they can do to help their clients achieve their overall rehabilitation goals.

We need to focus our efforts on community integration rather than on community maintenance. We already possess the strategies to decrease the number of days persons with psychiatric disabilities must spend in the hospital. People with severe psychiatric disabilities can be maintained in the community with community support and rehabilitation interventions, such as linking clients with paid companions or volunteers.

However, although it can be done, it does not follow that it is being done well or often enough. Community maintenance can be achieved for almost anyone, most of the time. It is yesterday's goal. The goal for the future is community integration; that is, can we help people who are maintained in the community function in the community more successfully and with greater satisfaction?

Consider the example of a man with a severe physical disability who has been equipped with a motorized wheelchair to help him live in the community. The question he might ask is, Now that I'm living in the community, where can I go in it? If architectural barriers limit his access to jobs, to school, to recreation, to transportation, then how has the wheelchair helped him become integrated?

The person with a psychiatric disability, who is being maintained in the community is in a similar predicament. The barriers are not architectural, but attitudinal and programmatic. The questions are the same, however: Now that I'm living in the community, where can I go in it? Where are the programs that reduce stigma? Where are the programs that help persons develop the skills and supports they need to be more integrated into the jobs, schools, and neighborhoods they want? The future challenge is how to help people function better in the community of their choice. The old goal of community maintenance is just that—yesterday's goal.

We need to think of psychiatric treatment and psychiatric rehabilitation not as competing models, but rather as complementary models. Proponents of psychiatric rehabilitation are sometimes perceived as being against medical treatment. This attitude has puzzled us, as we believe good treatment and good rehabilitation go hand in hand. Yet we hear from treatment professionals that the rehabilitation approach forces one to choose between rehabilitation and treatment.

It appears to us that this perceived antagonism may be caused by the disapproval rehabilitation professionals express about what they deem to be bad treatment—treatment that does not involve the clients in deciding their treatment, that uses medications over the long term without any understanding by or input of the client, that does not acknowledge or support the need for rehabilitation for persons who are severely psychiatrically disabled, and that does not recognize its own limitations (e.g., side effects and the obvious fact that medication can't teach skills or provide supports). Rehabilitation professionals are not against treatment per se.

Good treatment that involves the clients, that helps them to understand their treatment, and that recognizes and promotes the usefulness of rehabilitation is complementary to the psychiatric rehabilitation approach. Good treatment and psychiatric rehabilitation are a potent combination.

Vision of the Future

When we look into the future from the perspective of rehabilitation, we see, in general, an emerging consensus as to psychiatric rehabilitation philosophy, the continued expansion of psychosocial rehabilitation centers, the integration of the rehabilitation approach into existing mental health settings, a developing technology and research base, and increasing acknowledgment of the need for participation by consumers. We can also see several more specific developments.

First, we envision a mental health system in which mental health agencies employ a significant number of people who are current or former service recipients. Picture a state department of mental health's annual awards banquet. Picture the tables of current employees—former recipients of services seated next to career service providers, now joined by the common bond of working in a department—together trying to make life better for persons with psychiatric disabilities. Picture a mental health commissioner who has been a consumer of mental health services, or who is the parent of a current consumer.

This vision of the future has already arrived in some state divisions of vocational rehabilitation. The leadership and many of the divisions' employees are persons with physical disabilities. What today is a reality in some states would have been only a dream 30 years ago.

The field of psychiatric rehabilitation can work toward this vision by encouraging university training programs to seek and give preference to students recovering from a psychiatric disability. State departments of mental health can give preference to persons with psychiatric disabilities for hiring at all levels of the organization—professional, technical, clerical, and general help. State departments could actively seek supported employment slots within their settings. Mental health agencies, such as community mental health centers, community residential programs, and state

217

hospitals could do likewise. A farfetched vision? Only to the extent that no current steps are being taken toward it.

We envision a mental health system that believes that persons with psychiatric disabilities have the same aspirations and goals as anyone else. Consumers of mental health services, although rarely asked their goals by systems planners, can in fact articulate their goals. As might be expected, when asked about their goals, they mention the same goals as other persons— satisfying jobs, decent places to live, a chance to return to school, friendships, and a reduction of psychological distress.

During the past 5 years, we have attempted to learn about consumers' goals. Systematic assessments of various programs within the mental health service system (e.g., hospitals, partial-care settings, day treatment) have been conducted throughout the nation. These assessments routinely include interviews with consumers. Furthermore, we have trained hundreds of practitioners across North America during which we reviewed hundreds of audiotaped interviews in which persons with psychiatric disabilities discussed their goals with their practitioners. From these assessment interviews with consumers and in the tape recordings of clinical interviews, it was obvious that the goals of persons with psychiatric disabilities are very similar to the goals of everyone, disabled or not.

Concern has been expressed that persons with a severe psychiatric disability will not be able to set goals or that the goals they set will be destructive (e.g., want to harm someone) or unrealistic (e.g., want to be a symphony conductor). It has been our experience that, when given the opportunity, most persons with psychiatric disabilities can be helped to set reasonable goals. This is not to deny that at times persons with psychiatric disabilities will set unrealistic or even destructive goals. Rather, over time, as they become engaged in a process geared toward understanding what they want out of life rather than what their pathology has done to their life, their symptoms will become less dominating.

The majority of persons with psychiatric disabilities can be engaged in setting their living, learning, social, and working goals. The overall philosophy must be to give persons the opportunity to do so. When practitioners take the time to develop close

218

relationships with clients, they can facilitate an exploration of what the clients want. Practitioners must be as committed to listening to the goals of persons with psychiatric disabilities as psychopharmacologists are to assessing their symptoms.

We envision a mental health system in which the consumers play a major role in deciding what new programs are needed. If this vision became a reality, concepts such as *least restrictive environment* would be replaced by concepts such as *most preferred environment.* If consumers did play more of a role in these types of decisions, we would probably see different types of programs being developed. We would expect more consumer-operated, self-help programs, more educational and vocational programs for persons of normal intelligence rather than the traditional day treatment programs, more supported employment rather than sheltered workshops, more supported apartments rather than group homes.

The above are, of course, our guesses based on information from consumers. They need not be guesses. Policies could be set so that no major system change initiatives would occur without first surveying consumers as to what they think would be best. We assume we know, but we rarely ask. Asking the consumers of the service what they need and want should be a requirement.

We envision a mental health system that is driven by the clients' goals rather than by the system's goals. Not only are clients not asked how they would like to see services change, but their goals also do not drive the system. As alluded to previously, part of the problem in attaining this vision is that many people do not believe that persons with psychiatric disabilities have meaningful, realistic goals. Even if they do, however, system planners do not let these goals drive the system. Instead, agency and systemwide goals are stated (e.g., to provide comprehensive services or to contain costs). We have yet to hear clients state that their goal is to receive comprehensive services!

These goals and such others as to increase the number of psychosocial rehabilitation centers or to hire more case managers, are simply statements of objectives rather than statements of the desired outcomes. We hope the outcome for clients will change because of these objectives, but we cannot be sure. The system must be driven and held accountable based on the achievement of client goals, not on the implementation of system objectives.

We envision a mental health system that is committed to meeting the housing needs and preferences of persons with psychiatric disabilities. Attaining this vision would mean changing some current assumptions, for example, that clients can and should move along a continuum of housing options—not a particularly welcome idea especially if the persons being asked to move have just become successful and are satisfied in their current housing situations! Connecting a particular housing arrangement to participation in a particular service (e.g., day treatment) is also out-of-date. Services should be related to persons and not to their housing! Can persons with diabetes receive insulin only if they live in group homes? Can people who are blind have guide dogs only if they live in the country? Can persons with quadriplegia have a personal care attendant only if they live in a center for independent living?

Common sense, although admittedly not as common as it should be in the mental health field, must prevail as we move toward new housing alternatives. The concept of supported housing has been evolving out of the psychiatric rehabilitation approach. This vision of the future suggests that persons with psychiatric disabilities will be helped to live in the housing of their choice. Supports and services will be designed to help the person choose, get, and keep their preferred housing arrangement. The intensity, length, and type of services are a function of what the person needs and wants, not where the person lives.

We envision a mental health system in which practitioners can get excited about gradual, incremental changes in their clients and get rewarded for helping to achieve them. People with psychiatric disabilities often recover slowly, over a long time period. Seemingly small gains—a new skill learned, a new activity joined, smiling again—can be important benchmarks of progress.

Practitioners must be trained to look for and bring about these small changes, just as researchers must be trained to look for certain objects under a microscope. A trained eye sees and appreciates things under a microscope that an untrained eye cannot. So it is with psychiatric rehabilitation practitioners.

Quality assurance and management information systems must have the capacity to record subtle changes. The newer outcomes measures have started to reflect the legitimacy of incremental outcomes, such as part-time employment, supported employment, and improved levels of functioning. Measurement of outcome is not all or nothing.

We envision a day when the DSM *diagnosis will characterize schizophrenia not as an illness whose common course is "increasing deterioration between episodes," but as an impairment whose most common course is increased functioning over time.* As pointed out in chapter 2, it is instructive to look at how the prognosis for persons with developmental disability (e.g., Down's syndrome) has changed over time. At one time it was expected that most persons with such disabilities would live their lives in institutions, or certainly beyond the eyes of the community. Now, however, the prognosis for most persons with Down's syndrome is not lifelong institutionalization or total isolation from the community. Have the persons with Down's syndrome changed? Or has the way we treat persons with Down's syndrome and society's attitude toward them changed? Could not the same thing happen for persons with psychiatric disabilities? How much of their long-term prognosis is due to the impairment and how much is due to the lack of a rehabilitation-oriented service system, low expectations for recovery, and society's attitudes toward persons with psychiatric disabilities?

This vision of the future will be realized if we begin to make reasonable accommodations to impairments, not just through symptom management programs but with intensive rehabilitation programs. We must provide supportive environments that help persons cope in spite of impairments, and believe that a person's functioning can improve over the long term.

We envision a mental health system that does not define people who use the service by labels, but sees them first and foremost as people. The diagnostic labeling process sometimes results in persons with psychiatric disabilities incorporating the labels into their very beings. They consider themselves schizophrenic persons, paranoid, manic, low functioning, or worse yet, chronically mentally ill. They consider that all their reactions reflect this abnormality, and they lose sight of what a normal, typical reaction to a situation is.

The system that serves persons with impairments, and not impaired persons, must use language that emphasizes the person.

221

Rehabilitation programs can refer to their recipients of services as trainees, students, members, and so on. *Persons with schizophrenia* is the term preferred over referring to someone as a *schizophrenic*. Society and professionals alike must become sensitive to using terminology that does not emphasize the pathology. People are not impaired 24 hours a day, 365 days a year. In many ways they are more able than disabled. If we look at the whole person— the physical, intellectual, and spiritual aspects of the person as well as the emotional—we can see that the person with a psychiatric disability has a range of ideas, experiences, and beliefs.

We envision a mental health system that realizes that the outcomes desired by clients are impacted by the skills, knowledge, and attitudes of staff, not their academic credentials. People with psychiatric disabilities deserve to be helped by the persons most qualified to help. However, the most qualified are not necessarily the most credentialed professionals. What makes people qualified are their psychiatric rehabilitation skills, knowledge, and attitudes.

When recovering clients are asked to reflect on what has been most helpful to them and why, they usually mention characteristics of the staff. Most important to them are the skills of the staff, their ability to connect, teach, and be supportive. The attitudes of staff reflect caring, respect, and a belief in the capacity of the person to change. These human attributes seem much more important than do credentials.

The mental health system must realize that these human dimensions are more critical and important than credentials. In psychiatric rehabilitation neither staff labels nor client labels are of primary importance. Advancement in the field needs to be a function of what practitioners can do and not what diploma they have. Universities must ensure that graduates of their training programs, in addition to the unique contributions of their discipline, can connect, teach, and support in the context of a caring and hopeful attitude. Persons with these characteristics, no matter what their credentials, are in fact the most qualified to serve.

We see a mental health system that creates new programs rather than new labels for clients and their families. Such labels as the young adult chronic, the chronically mentally ill, the most at risk client, the high EE family member, the unmotivated client categorize persons by their deficits. Although the developers of the labels have tried to call attention to a specific group of people who lack appropriate services, use of the labels has been at times unfortunate. The designation of the label has often been followed by lowered expectations for persons, that is, not much can be done because they are chronic or unmotivated. The clients get blamed and are considered to be chronic or unmotivated when the problem might be that the available program is unappealing and ineffective. Clients are being blamed for not benefiting from programs because they are too withdrawn, too lethargic, or too unpredictable. This is as meaningful as blaming persons who are blind for being difficult to rehabilitate because they cannot see!

The appearance of new labels is a portent that new services must be developed. The direction is service system change—not more definitive labeling.

We envision a system in which services of state divisions of vocational rehabilitation (VR) are an entitlement for persons with psychiatric disabilities rather than services for which they must qualify. Severe psychiatric disability in itself should entitle people for VR services. The research clearly shows how difficult it is to predict whether a person will profit from vocational rehabilitation services. It is possible that a person who did not profit from vocational rehabilitation services the first time will profit from them the second or third time. The task of determining beforehand whether vocational rehabilitation services will help a person with a severe psychiatric disability is practically impossible, and the policy and procedures of the VR system need to acknowledge that fact.

A prejudiced society is a great barrier to the vocational rehabilitation of persons with psychiatric disabilities. VR systems certainly need not erect additional ones.

We envision a mental health system committed to helping persons with psychiatric disabilities achieve their residential and vocational goals by using psychiatric rehabilitation technology. Currently, some people mistakenly assume that the reasons persons with psychiatric disabilities do not achieve greater success is because they cannot make realistic choices or because their symptoms will always interfere with their lives. In reality, however, persons with psychiatric disabilities can make more realistic choices when given more experiences from which they can develop goals. For example, in the vocational area, transitional employment placements, volunteer jobs, clubhouse positions, vocational internships, and job shadowing experiences give people with psychiatric disabilities the experiential base from which to make their vocational decisions.

Persons with psychiatric disabilities lack the experiences most other people have. Their young adulthood often was consumed by the psychiatric impairment and the treatment. They did not develop educationally or vocationally as did their peers. Thus, people with psychiatric disabilities may be considered immature or inexperienced. They often learned the career of mental patient rather than more productive careers.

In order to develop new careers, persons with psychiatric disabilities need to learn or relearn different skills and gain access to new supports. Skills and supports, not symptoms, correlate with rehabilitation outcome. Yesterday's technology is not enough. New technology needs to be used (e.g., functional assessment, skills teaching, and case management).

Just within the last decade, the mental health service system adopted improved residential and vocational functioning as important client goals. Once committed to these goals, an ever-expanding technology should emerge to help increasing numbers of persons with psychiatric disabilities resume the pursuit of their residential and vocational goals.

We envision a mental health system that denies no one the opportunity for rehabilitation interventions, for as many times and for as long as they are needed and wanted. The support component of a rehabilitation intervention is often needed on a continual basis. Do we ask people to turn in their wheelchairs after one year? Are arbitrary time limits set on how long a person can have a guide dog? In psychiatric rehabilitation the support component needs to be available on an as-needed basis for as long as the support is necessary to help a person function. An arbitrary time period should not be set for support. Often the support might be needed only intermittently, perhaps at times of crises or new opportunities. Persons with psychiatric disabilities need to know that the support is there, if needed.

224

The concepts of supported housing, supported education, and supported employment recognize the importance of ongoing support. The community support-system initiative stresses the value of ongoing support. In physical medicine and rehabilitation the need for such support is obvious, and its arbitrary withdrawal would border on the criminal. Proponents of psychiatric rehabilitation must ensure that persons with a need for psychiatric rehabilitation receive the rehabilitation interventions they need, as often and for as long as they need and want them.

We envision a mental health system in which persons with psychiatric disabilities can receive the help they need and want without having to pay the ultimate price—their dignity. The extent to which some people will resist would-be helpers who do not respect their dignity is truly amazing, sometimes bordering on the superhuman. Some homeless persons with psychiatric disabilities exemplify that phenomenon as do some persons who resist treatment or dropout. The question becomes, Must we demean or infantilize people, or strip away their humanity as part of the helping process? People who know they need help will still refuse it if accepting that help demeans them in the process.

Persons with psychiatric disabilities who need help but actively resist that help may be saying more about the quality of that help than they are saying about themselves. For some persons it may not be their ''craziness'' that is causing them to resist help, but rather their humanness.

To achieve our vision, we must examine how we provide help. Is the help offered in a way that respects the person's dignity? How many dropouts can become drop-ins, how many involuntaries can become voluntary, if the helping system respects their dignity?

We envision a mental health service system that recognizes and appreciates the positive attributes of persons with psychiatric disabilities. The mental health field has tended to focus on negative traits when referring to its clientele. In particular, young adults with psychiatric disabilities have been characterized in the literature as disturbed, disabled, drugged, or young adult chronics. Their many positive attributes have not been acknowledged.

Other groups of person with disabilities now receive much

more favorable attention, both by the press and the professionals who provide the services. For example, consider the following description of a group of disabled persons:

> The majority of them are unemployed; they have a higher suicide rate; many of them are supported by SSI or SSDI; they are frequently hospitalized; they are a significant cost and burden to society; some even neglect their own self-care; others are angry at the barriers that society has used to isolate them.

Who are these people? They are persons with *physical* disabilities. Fortunately, society and the persons that help them do not emphasize these traits. Rather, they emphasize how they are meeting challenges, overcoming unfair attitudes, and getting on with their lives. We need to start stressing the positive attributes of persons with psychiatric disabilities. They may be above normal in intelligence or have achieved an above-average education; some may be creative and talented; others may have high career aspirations or unique senses of humor. In short, they are often worth bragging about.

Stigma in society cannot be reduced if it emanates from the professionals who should be helping to eradicate it. Shouldn't mental health professionals model the behaviors we wish society to emulate? It starts with the professionals stressing the positive, and then the media, policy makers, elected officials, and business people will follow suit.

We envision a society that places a greater priority on life enhancement than on cost containment for persons with psychiatric disabilities. Persons with psychiatric disabilities need not have dollar signs attached to their rights. Do rehabilitation services for them have to return more dollars to the treasury than the services cost? We think not. The opportunity for greater independence for all persons with disabilities must become a right, not a privilege.

We do not seem as worried about the cost to society for persons with physical illnesses who need expensive treatment, and especially for children who need interventions of heroic proportions. These charitable attitudes are examples of society's finest moments.

Yet when proposals for new and effective interventions are suggested for persons with psychiatric disabilities, sometimes the first question asked is, How much will it cost? Often this question comes from people who might be expected to advocate better programs—family members and professionals! Don't we believe the intervention will really work or the clients merit the attempt? If we, who should be concerned with life enhancement, worry first about cost containment, then who will advocate life enhancement?

Both the Education for All Handicapped Children Act and the Rehabilitation Act were advocated on the basis of life enhancement. The laws' removal of architectural and educational barriers has been a notable achievement. Are any of the advocates for these laws apologizing for the cost? We think not. Instead they are advocating for more improvement and enforcement of the laws.

There is a natural order with respect to life enhancement and cost containment. First, we develop and implement programs that do a better job of enhancing the lives of persons with psychiatric disabilities. Then we make them as inexpensive as possible. Life enhancement precedes cost containment.

We envision a world in which people are attitudinally accessible. One of the greatest barriers to the clients' achievement of their rehabilitation goals is the society in which they have to achieve these goals. What often stands in the way of a suitable job or a decent place to live is not an inaccessible building but an inaccessible person. Employers, landlords, teachers, and neighbors possess the capacity to unleash the talents of persons with psychiatric disabilities—if only they would unharness themselves from their prejudices. We wonder how many more clients could be employed, return to school, live more independently—without the clients themselves even changing—if only the attitudes change in the world.

If this vision is ever to be realized, we must relentlessly advocate treating clients like other persons. This starts with how professionals treat their clients. For if we treat people as if they are schizophrenics, manics, and paranoids rather than people, how can we expect society to interact with persons rather than a stereotypical image of a disease.

227

Concluding Comment

To speak of an ideal world is stimulating, a luxury we should all experience. How easy it is to dream, to hope. But, as Benjamin Franklin wrote, "If we live by hope alone we will probably die fasting." So where do we go with our vision? This is not merely an exercise. It describes the larger and more idealistic mission toward which we work. To modify slightly the words of Chesterton, "It's not that the ideal has been tried and found wanting, it's that the ideal has been found difficult and left untried."

The key to coming closer to one's vision is action. The price of a vision, the fee for dreaming, is *acting* on one's hopes and dreams. We must follow through and do the hard work that is both stimulated and made easier by our dreams. Dreams without action can be deceiving; action without dreams can be erratic and mindless.

Achieving the vision of psychiatric rehabilitation depends on its growing knowledge base, philosophy, and the excellence of its technology. Achieving the vision also depends on the people who practice in the psychiatric rehabilitation field, the programs in which they practice, and the systems that support their practice. If the vision does in fact come closer to reality, then the lives of people who have experienced psychiatric disabilities should be the better for it—truly a worthwhile enterprise.

References

Adler, D. A., Drake, R. E., Berlant, J., Ellison, J. M., & Carson, D. (1987). Treatment of the nonpsychotic chronic patient: A problem of interactive fit. *American Journal of Orthopsychiatry*, *57*, 579–586.

Alevizos, P., & Callahan, E. (1977). The assessment of psychotic behavior. In A. Ciminero, K. Calhoun, & H. Adams (Eds.), *Handbook of behavioral assessment* (pp. 683–721). New York: John Wiley & Sons.

American Psychiatric Association. (1987). *Diagnostic and statistical manual for mental disorders* (3rd ed., rev.). Washington, DC: Author.

American Psychiatric Association. (1980). *Diagnostic and statistical manual of mental disorders* (3rd ed.). Washington, DC: Author.

Anderson, C. M., Hogarty, G. E., & Reiss, D. J. (1980). Family treatment of adult schizophrenic patients: A psychoeducational approach. *Schizophrenia Bulletin*, *6*, 490–505.

Anderson, C. M., Hogarty, G., & Reiss, D. J. (1981). The psychoeducational family treatment of schizophrenia. In M. Goldstein (Ed.), *New developments in interventions with families of schizophrenics* (New Directions for Mental Health Services, No. 12). San Francisco: Jossey-Bass.

Angelini, D., Potthof, P., & Goldblatt, R. (1980). *Multi-Functional Assessment Instrument*. Unpublished manuscript, Rhode Island Division of Mental Health, Cranston, RI.

Anthony, W. A. (1972). Societal rehabilitation: Changing society's attitudes toward the physically and mentally disabled. *Rehabilitation Psychology*, *19*, 117–126.

229

Anthony, W. A. (1977). Psychological rehabilitation: A concept in need of a method. *American Psychologist, 32*, 658–662.

Anthony, W. A. (1979). *The principles of psychiatric rehabilitation.* Baltimore: University Park Press.

Anthony, W. A. (Ed.). (1980). Rehabilitating the person with a psychiatric disability: The state of the art [Special issue]. *Rehabilitation Counseling Bulletin, 24.*

Anthony, W. A. (1982). Explaining "psychiatric rehabilitation" by an analogy to "physical rehabilitation." *Psychosocial Rehabilitation Journal, 5*(1), 61-65.

Anthony, W. A. (1984). The one-two-three of client evaluation in psychiatric rehabilitation settings. *Psychosocial Rehabilitation Journal, 8*(2), 85–87.

Anthony, W. A., & Blanch, A. K. (1989). Research ·on community support services: What have we learned? *Psychosocial Rehabilitation Journal, 12*(3), 55–81.

Anthony, W. A., Buell, G. J., Sharratt, S., & Althoff, M. E. (1972). Efficacy of psychiatric rehabilitation. *Psychological Bulletin, 78*, 447–456.

Anthony, W. A., & Carkhuff, R. R. (1976). *The art of health care: A handbook of psychological first aid skills.* Amherst, MA: Human Resource Development Press.

Anthony, W. A., & Carkhuff. R. R. (1978). The functional professional therapeutic agent. In A. Gurman & A. Razin (Eds.), *Effective psychotherapy* (pp. 84–119). London: Pergamon Press.

Anthony, W. A., Cohen, M. R., & Cohen, B. F. (1983). Philosophy, treatment process, and principles of the psychiatric rehabilitation approach. In L. L. Bachrach (Ed.), *Deinstitutionalization* (New Directions for Mental Health Services, No. 17, pp. 67–69). San Francisco: Jossey-Bass.

Anthony, W. A., Cohen, M. R., & Cohen, B. F. (1984). Psychiatric rehabilitation. In J. A. Talbott (Ed.), *The chronic mental patient: Five years later* (pp. 137–157). Orlando, FL: Grune & Stratton.

Anthony, W. A., Cohen, M. R., & Farkas, M. D. (1982). A psychiatric rehabilitation treatment program: Can I recognize one if I see one? *Community Mental Health Journal, 18*, 83–96.

Anthony, W. A., Cohen, M. R., & Farkas, M. D. (1987). Training and technical assistance in psychiatric rehabilitation. In A. T. Meyerson & T. Fine (Eds.), *Psychiatric disability: Clinical, legal, and administrative dimensions* (pp. 251–269). Washington, DC: American Psychiatric Press.

Anthony, W. A., Cohen, M. R., & Farkas, M. D. (1988). Professional pre-service training for working with the long-term mentally ill. *Community Mental Health Journal, 24*, 258–269.

Anthony, W. A., Cohen, M. R., Farkas, M. D., & Cohen, B. F. (1988). Clinical care update: Case management—more than a response to a dysfunctional system. *Community Mental Health Journal, 24*, 219–228.

Anthony, W. A., Cohen, M. R., & Kennard, W. A. (1989). *Understanding the current facts and principles of mental health system planning.* Unpublished manuscript, Boston University, Center for Psychiatric Rehabilitation, Boston.

Anthony, W. A., Cohen, M. R., & Nemec, P. B. (1987). Assessment in psychiatric rehabilitation. In B. Bolton (Ed.), *Handbook of measurement and evaluation in rehabilitation* (pp. 299–312). Baltimore: Paul Brookes.

Anthony, W. A., Cohen, M. R., & Pierce, R. M. (1980). *Instructor's guide to the psychiatric rehabilitation practice series.* Baltimore: University Park Press.

Anthony, W. A., Cohen, M. R., & Vitalo, R. L. (1978). The measurement of rehabilitation outcome. *Schizophrenia Bulletin, 4,* 365–383.

Anthony, W. A., & Farkas, M. D. (1982). A client outcome planning model for assessing psychiatric rehabilitation interventions. *Schizophrenia Bulletin, 8,* 13–38.

Anthony, W. A., & Farkas, M. D. (1989). The future of psychiatric rehabilitation. In M. D. Farkas & W. A. Anthony (Eds.), *Psychiatric rehabilitation programs: Putting theory into practice* (pp. 226–239). Baltimore: Johns Hopkins University Press.

Anthony, W. A., Howell, J., & Danley, K. S. (1984). Vocational rehabilitation of the psychiatrically disabled. In M. Mirabi (Ed.), *The chronically mentally ill: Research and services* (pp. 215–237). Jamaica, NY: Spectrum Publications.

Anthony, W. A., & Jansen, M. A. (1984). Predicting the vocational capacity of the chronically mentally ill: Research and policy implications. *American Psychologist, 39,* 537–544.

Anthony, W. A., Kennard, W. A., O'Brien, W., & Forbess, R. (1986). Psychiatric rehabilitation: Past myths and current realities. *Community Mental Health Journal, 22,* 249–264.

Anthony, W. A., & Liberman, R. P. (1986). The practice of psychiatric rehabilitation: Historical, conceptual, and research base. *Schizophrenia Bulletin, 12,* 542–559.

Anthony, W. A., & Margules, A. (1974). Toward improving the efficacy of psychiatric rehabilitation: A skills training approach. *Rehabilitation Psychology, 21,* 101–105.

Anthony, W. A., & Nemec, P. B. (1984). Psychiatric rehabilitation. In A. S. Bellack (Ed.), *Schizophrenia: Treatment, management, and rehabilitation* (pp. 375–413). Orlando, FL: Grune & Stratton.

Anthony, W. A., & Stroul, B. (1986). *The community support system: An idea whose time has come—and stayed.* Unpublished manuscript, Boston University, Center for Psychiatric Rehabilitation, Boston.

Appleton, W. (1974). Mistreatment of patients' families by psychiatrists. *American Journal of Psychiatry, 131,* 655–657.

Armstrong, B. (1977) A federal study of deinstitutionalization: How the government impedes its goal. *Hospital and Community Psychiatry, 28,* 417, 425.

Armstrong, H. E., Rainwater, G., & Smith, W. R. (1981). Student-like behavior as a function of contingent social interaction in a psychiatric day treatment program. *Psychological Reports, 48,* 495–500.

Arthur, G., Ellsworth, R. B., & Kroeker, D. (1968). Schizophrenic patient post-hospital community adjustment and readmission. *Social Work, 13,* 78–84.

Aspy, D. (1973). *Toward a technology for humanizing education.* Champaign, IL: Research Press.

Aspy, D., & Roebuck, F. (1977). *Kids don't learn from people they don't like.* Amherst, MA: Human Resource Development Press.

Aveni, C. A., & Upper, D. (1976, May). *Training psychiatric patients for community living.* Paper presented at the meeting of the Midwestern Association of Behavior Analysis, Chicago.

Avison, W. R., & Speechley, K. N. (1987). The discharged psychiatric patient: A review of social, social-psychological, and psychiatric correlates of outcome. *American Journal of Psychiatry, 144,* 10–18.

Ayd, F. (1974). Treatment resistant patients: A moral, legal and therapeutic challenge. In F. Ayd (Ed.), *Rational psychopharmacotherapy and the right to treatment.* Baltimore: Ayd Medical Communications.

Azrin, N., & Philip, R. (1979). The job club method for the job handicapped: A comparative outcome study. *Rehabilitation Counseling Bulletin,* December, 144–156.

Bachrach, L. L. (1976a). A note on some recent studies of mental hospital patients released into the community. *American Journal of Psychiatry, 133,* 73–75.

Bachrach, L. L. (1976b). *Deinstitutionalization: An analytical review and sociological perspective.* Rockville, MD: National Institute of Mental Health.

Bachrach, L. L. (1980). Overview: Model programs for chronic mental patients. *American Journal of Psychiatry, 137,* 1023–1036.

Bachrach, L. L. (1982a). Assessment of outcomes in community support systems: Results, problems, and limitations. *Schizophrenia Bulletin, 8,* 39–60.

Bachrach, L. L. (1982b). Program planning for young adult chronic patients. In B. Pepper & H. Ryglewicz (Eds.), *The young adult chronic patient* (New Directions for Mental Health Services, No. 14, p. 254). San Francisco: Jossey-Bass.

Bachrach, L. L. (1983). New directions in deinstitutionalization planning. *Deinstitutionalization* (New Directions for Mental Health Services, No. 17, pp. 3–106). San Francisco: Jossey-Bass.

Bachrach, L. L. (1986a). Deinstitutionalization: What do the numbers mean? *Hospital and Community Psychiatry, 37,* 118–121.

Bachrach, L. L. (1986b). The future of the state mental hospital. *Hospital and Community Psychiatry, 37,* 467–474.

Bachrach, L. L. (1988a). Defining chronic mental illness: A concept paper. *Hospital and Community Psychiatry*, *39*, 383–388.

Bachrach, L. L. (1988b). On exporting and importing model programs. *Hospital and Community Psychiatry*, *39*, 1257–1258.

Bachrach, L. L. (1989). The legacy of model programs. *Hospital and Community Psychiatry*, *40*, 234–235.

Bachrach, L. L. (in press). The context of care for the chronic mental patient with substance abuse problems. *Psychiatric Quarterly*.

Baker, F., Kazarian, S. S., Helmes, E., Ruckman, M., & Tower, N. (1987). Perceived attitudes of schizophrenic inpatients in relation to rehospitalization. *Journal of Consulting and Clinical Psychology*, *55*, 775-777.

Baker, F., & Weiss, R. S. (1984). The nature of case manager support. *Hospital and Community Psychiatry*, *35*, 925–928.

Ballantyne, R. (1983). Community rehabilitation services: A new approach to aftercare. *Network*, *3*, 4–6.

Barbee, M. S., Berry, K. L., & Micek, L. A. (1969). Relationship of work therapy to psychiatric length of stay and readmission. *Journal of Consulting and Clinical Psychology*, *33*, 735–738.

Barofsky, I., & Connelly, C. E. (1983). Problems in providing effective care for the chronic psychiatric patient. In I. Barofsky & R. D. Budson (Eds.), *The chronic psychiatric patient in the community* (pp. 83–129). New York: SP Medical and Scientific Books.

Barter, J. T. (1983). California: Transformation of mental health care: 1957-1982. In J. A. Talbott (Ed.), *Unified health services: Utopia unrealized* (New Directions in Mental Health Services, No. 18, pp. 7–18). San Francisco: Jossey-Bass.

Barton, W. E., & Barton, G. M. (1983). *Mental health administration: Principles and practices* (Vols. 1 & 2). New York: Human Sciences Press.

Bassuk, E. L., & Gerson, S. (1978). Deinstitutionalization and mental health services. *Scientific American*, *238*(2), 46–53.

Beard, J. H., Malamud, T. J., & Rossman, E. (1978). Psychiatric rehabilitation and long term rehospitalization rates: The findings of two research studies. *Schizophrenia Bulletin*, *4*, 622–635.

Beard, J. H., Pitt, R. B., Fisher, S. H., & Goertzel, V. (1963). Evaluating the effectiveness of a psychiatric rehabilitation program. *American Journal of Orthopsychiatry*, *33*, 701–712.

Beard, J. H., Propst, R. N., & Malamud, T. J. (1982). The Fountain House model of psychiatric rehabilitation. *Psychosocial Rehabilitation Journal*, *5*(1), 47–53.

Becker, P., & Bayer, C. (1975). Preparing chronic patients for community placement: A four-stage treatment program. *Hospital and Community Psychiatry*, *26*, 448–450.

233

Bell, R. L. (1970). Practical applications of psychodrama: Systematic role playing teaches social skills. *Hospital and Community Psychiatry*, *21*, 189-191.

Bellack, A. S., Morrison, R. L., & Mueser, K. T. (1989). Social problem solving in schizophrenia. *Schizophrenia Bulletin*, *15*, 101–116.

Berzinz, J. I., Bednar, R. L., & Severy, L. J. (1975). The problem of intersource consensus in measuring therapeutic outcomes: New data and multivariate perspectives. *Journal of Clinical Psychology*, *84*(1), 10–19.

Berzon, P., & Lowenstein, B. (1984). A flexible model of case management. In B. Pepper & H. Ryglewicz (Eds.), *Advances in treating the young adult chronic patient* (New Directions for Mental Health Services, No. 21, pp. 49- 57). San Francisco: Jossey-Bass.

Bevilacqua, J. J. (1984). *Chronic mental illness, a problem in politics.* Paper presented at the National Conference on the Chronic Mental Patient, Kansas City, KS, August 3.

Bigelow, D., & Young, D. (1983). *Effectiveness of a case management program.* Unpublished manuscript, University of Washington, Graduate School of Nursing, Seattle.

Blackman, S. (1982). Paraprofessional and patient assessment criteria of patient's recovery: Why the discrepancy? *Journal of Clinical Psychology*, *37*(4), 903–907.

Bolton, B. A. (1974). A factor analysis of personal adjustment and vocational measures of client change. *Rehabilitation Counseling Bulletin*, *18*, 99–104.

Bolton, B. (1978). Dimensions of client change: A replication. *Rehabilitation Counseling Bulletin*, *22*, 8–14.

Bond, G. R. (1984). An economic analysis of psychosocial rehabilitation. *Hospital and Community Psychiatry*, *35*, 356–362.

Bond, G. R., & Dincin, J. (1986). Accelerating entry into transitional employment in a psychosocial rehabilitation agency. *Rehabilitation Psychology*, *31*, 143–154.

Bond, G. R., Dincin, J., Setze, P. J., & Witheridge, T. F. (1984). The effectiveness of psychiatric rehabilitation: A summary of research at Thresholds. *Psychosocial Rehabilitation Journal*, *7*(4), 6–22.

Bond, G. R., & Friedmeyer, M. H. (1987). Predictive validity of situational assessment at a psychiatric rehabilitation center. *Rehabilitation Psychology*, *32*, 99–112.

Bond, G. R., Miller, L. D., Krumwied, R. D., & Ward, R. S. (1988). Assertive case management in three CMHCs: A controlled study. *Hospital and Community Psychiatry*, *39*, 411–418.

Bond, G. R., Witheridge, T. F., Wasmer, D., Dincin, J., McRae, S. A., Mayes, J., & Ward, R. S. (1989). Short-term assertive outreach and service coordination: A comparison of two crisis housing alternatives to psychiatric hospitalization. *Hospital and Community Psychiatry, 40*, 177–183.

Borland, A., McRae, J., & Lycan, C. (1989). Outcomes of five years of continuous intensive case management. *Hospital and Community Psychiatry, 40*, 369–376.

Borys, S., & Fishbein, S. M. (1983). *Partial care technical assistance project: Pretest results* (Research and Evaluation Report). Trenton: New Jersey Division of Mental Health and Hospitals.

Bowker, J. P.(Ed.). (1985). *Education for practice with the chronically mentally ill: What works?* Washington, DC: Council on Social Work Education.

Brady, J. P. (1984). Social skills training for psychiatric patients: II. Clinical outcome studies. *American Journal of Psychiatry, 141*, 491–498.

Braun, P., Kochansky, G., Shapiro, R., Greenberg, S., Gudeman, J. E., Johnson, S., & Shore, M. F. (1981). Overview: Deinstitutionalization of psychiatric patients: A critical review of outcome studies. *American Journal of Psychiatry, 138*, 736–749.

Breier, A., & Strauss, J. S. (1983). Self-control in psychiatric disorders. *Archives of General Psychiatry, 40*, 1141–1145.

Brekke, J. S., & Test, M. A. (1987). An empirical analysis of services delivered in a model community support program. *Psychosocial Rehabilitation Journal, 10*(4), 51–61.

Brook, S., Fantopoulos, I., Johnston, F., & Goering, P. (1989). Training volunteers to work with the chronically mentally ill in the community. *Hospital and Community Psychiatry, 40*, 853–855.

Brooks, G. W. (1981). Vocational rehabilitation. In J. A. Talbott (Ed.), *The chronic mentally ill* (pp. 96–100). New York: Human Sciences Press.

Brown, G. W., Birley, J. L. T., & Wing, J. K. (1972). Influence of family life on the course of schizophrenic disorders: A replication. *British Journal of Psychiatry, 121*, 241–258.

Brown, M. A., & Basel, D. (1989). A five-stage vocational rehabilitation program: Laurel Hill Center, Eugene, Oregon. In M. D. Farkas & W. A. Anthony (Eds.), *Psychiatric rehabilitation programs: Putting theory into practice* (pp. 108–116). Baltimore: Johns Hopkins University Press.

Brown, P. (1982). Approaches to evaluating the outcome of deinstitutionalization: A reply to Christenfeld. *Journal of Psychology, 10*, 276–281.

Burns, B. J., Burke, J. D., & Kessler, L. G. (1981). Promoting health-mental health coordination: Federal efforts. In A. Broskowski, E. Marks, & S. H. Budman (Eds.), *Linking health and mental health*. Beverly Hills, CA: Sage Publications.

Byalin, K., Jed, J., & Lehman, S. (1982). *Family intervention with treatment-refractory chronic schizophrenics*. Paper presented at 20th International Congress of Applied Psychology, Edinburgh, Scotland.

Cannady, D. (1982). Chronics and cleaning ladies. *Psychosocial Rehabilitation Journal, 5*(1), 13–16.

Caplan, N. (1980). What do we know about knowledge utilization? In L. A. Braskamp & R. D. Brown (Eds.), *New directions for program education* (No. 5). San Francisco: Jossey-Bass.

Caragonne, P. (1981). An analysis of the function of the case manager in four mental health social service settings (Doctoral dissertation, University of Michigan, 1981). *Dissertation Abstracts International, 41*(7), 3262A.

Caragonne, P. (1983). *A comparison of case management work activity and current models of work activity within the Texas Department of Mental Health and Mental Retardation* (Report for the Texas Department of Mental Health and Mental Retardation). Austin, TX: Texas Department of Mental Health and Mental Retardation.

Carkhuff, R. R. (1968). The differential functioning of lay and professional helpers. *Journal of Counseling Psychology, 15*, 117–126.

Carkhuff, R. R. (1969). *Helping and human relations* (Vols. 1 & 2). New York: Holt, Rinehart & Winston.

Carkhuff, R. R. (1971). *The development of human resources*. New York: Holt, Rinehart & Winston.

Carkhuff, R. R. (1974). *The art of problem-solving*. Amherst, MA: Human Resource Development Press.

Carkhuff, R. R., & Anthony, W. A. (1979). *The skills of helping: An introduction to counseling skills*. Amherst, MA: Human Resource Development Press.

Carkhuff, R. R., & Berenson, B. G. (1976). *Teaching as treatment: An introduction to counseling and psychotherapy*. Amherst, MA: Human Resource Development Press.

Carling, P. J., & Broskowski, A. (1986). Psychosocial rehabilitation programs as a challenge and an opportunity for community mental health centers. *Psychosocial Rehabilitation Journal, 10*(1), 39–48.

Carling, P. J., Miller, S., Daniels, L. V., & Randolph, F. L. (1987). A state mental health system with no state hospital: The Vermont feasibility study. *Hospital and Community Psychiatry, 38*, 617–624.

Carpenter, W. T. (1979). Clinical research methods applicable to the study of treatment effects in chronic schizophrenic patients. In C. F. Bater & T. Melnechuk (Eds.), *Perspectives in schizophrenia research: Presentations and sessions of the VA Advisory Conference on Chronic Schizophrenia* (pp. 297-311). New York: Raven Press.

Carpenter, W. T., Heinrichs, D. W., & Hanlon, T. E. (1987). A comparative trial of pharmacologic strategies in schizophrenia. *American Journal of Psychiatry, 144*, 1466–1470.

Carpenter, W. T., McGlashan, T. H., & Strauss, J. S. (1977). The treatment of acute schizophrenia without drugs: An investigation of some current assumptions. *American Journal of Psychiatry, 134*, 14–20.

Castaneda, D., & Sommer, R. (1986). Patient housing options as viewed by parents of the mentally ill. *Hospital and Community Psychiatry, 37*, 1238-1242.

Center for Psychiatric Rehabilitation (1984). *Annual report for the National Institute of Handicapped Research.* Boston: Boston University.

Center for Psychiatric Rehabilitation (1989). *Research and training center final report (1984–1989).* Boston: Boston University.

Center for Psychiatric Rehabilitation Staff (1989). Refocusing on locus. *Hospital and Community Psychiatry, 40,* 418.

Chamberlin, J. (1978). *On our own: Patient-controlled alternatives to the mental health system.* New York: Hawthorn Books.

Chamberlin, J. (1984). Speaking for ourselves. An overview of the ex- psychiatric inmates' movement. *Psychosocial Rehabilitation Journal, 8*(2), 56–63.

Chamberlin, J. (1989). Ex-patient groups and psychiatric rehabilitation. In M. D. Farkas & W. A. Anthony (Eds.), *Psychiatric rehabilitation programs: Putting theory into practice* (pp. 207–216). Baltimore: Johns Hopkins University Press.

Cheadle, A. J., Cushing, D., Drew, C., & Morgan, R. (1967). The measurement of the work performance of psychiatric patients. *British Journal of Psychiatry, 113,* 841–846.

Cheadle, A. J., & Morgan, R. (1972). The measurement of work performance of psychiatric patients: A reappraisal. *British Journal of Psychiatry, 120,* 437–441.

Cheek, F. E., & Mendelson, M. (1973). Developing behavior modification programs with an emphasis on self control. *Hospital and Community Psychiatry, 24,* 410–416.

Ciminero, A., Calhoun, K., & Adams, H. (1977). *Handbook of behavioral assessment.* New York: John Wiley & Sons.

Cnaan, R. A., Blankertz, L., Messinger, K. W., & Gardner, J. R. (1988). Psychosocial rehabilitation: Toward a definition. *Psychosocial Rehabilitation Journal, 11*(4), 61–77.

Cohen, B. F., & Anthony, W. A. (1984). Functional assessment in psychiatric rehabilitation. In A. S. Halpern & M. J. Fuhrer (Eds.), *Functional assessment in rehabilitation* (pp. 79–100). Baltimore: Paul Brookes.

Cohen, B. F., Ridley, D. E., & Cohen, M. R. (1985). Teaching skills to severely psychiatrically disabled persons. In H. A. Marlowe & R. B. Weinberg (Eds.), *Competence development: Theory and practice in special populations* (pp. 118-145). Springfield, IL: Charles C Thomas.

Cohen, M. R. (1981). *Improving interagency collaboration between vocational rehabilitation and mental health agencies: A conference summary report* (Report). Boston: Boston University, Center for Psychiatric Rehabilitation.

Cohen, M. R. (1985). *Training professionals for work with persons with long-term mental illness.* Presentation at the CSP Project Director's meeting, Oct. 16–18, Chicago, IL.

237

Cohen, M. R. (1989). Integrating psychiatric rehabilitation into mental health systems. In M. D. Farkas & W. A. Anthony (Eds.), *Psychiatric rehabilitation programs: Putting theory into practice* (pp. 162–170, 188–191). Baltimore: Johns Hopkins University Press.

Cohen, M. R., & Anthony, W. A. (1988). A commentary on planning a service system for persons who are severely mentally ill: Avoiding the pitfalls of the past. *Psychosocial Rehabilitation Journal, 12*(1), 69–72.

Cohen, M. R., Danley, K. S., & Nemec, P. B. (1985). *Psychiatric rehabilitation training technology: Direct skills teaching* (Trainer package). Boston: Boston University, Center for Psychiatric Rehabilitation.

Cohen, M. R., Farkas, M. D., & Cohen, B. F. (1986). *Psychiatric rehabilitation training technology: Functional assessment* (Trainer package). Boston: Boston University, Center for Psychiatric Rehabilitation.

Cohen, M. R., Farkas, M. D., Cohen, B. F., & Unger, K. V. (1990). *Psychiatric rehabilitation training technology: Setting an overall rehabilitation goal* (Trainer package). Boston: Boston University, Center for Psychiatric Rehabilitation.

Cohen, M. R., Nemec, P. B., Farkas, M. D., & Forbess, R. (1989). *Psychiatric rehabilitation training technology: Case management* (Trainer package). Boston: Boston University, Center for Psychiatric Rehabilitation.

Cohen, M. R., Vitalo, R. L., Anthony, W. A., & Pierce, R. M. (1980). *The psychiatric rehabilitation practice series: Book 6. The skills of community service coordination.* Baltimore: University Park Press.

Connors, K. A., Graham, R. S., & Pulso, R. (1987). Playing store: Where is the vocational in psychiatric rehabilitation? *Psychosocial Rehabilitation Journal, 10*(3), 21–33.

Connors, K. A., Graham, R. S., & Pulso, R. (1987). Playing store: Where is the vocational in psychiatric rehabilitation? *Psychosocial Rehabilitation Journal, 5*(1), 35–39.

Cook, D. W. (1983, April). The accuracy of work evaluator and client predictions of client vocational competency and rehabilitation outcome. *Journal of Rehabilitation,* 46–48.

Cornhill Associates. (1980). *Needs Assessment Instrument.* Unpublished manuscript, Newton, MA.

COSMOS (1988). *Mental health planning news, 3*(1). Washington, DC: National Technical Assistance Center for Mental Health Planning.

Cozby, P. C. (1989). *Methods in behavioral research* (4th ed.). Mountain View, CA: Mayfield.

Craig, T. J. H., Peer, S. M., & Ross, M. D. (1989). Psychiatric rehabilitation in a state hospital transitional residence: The Cottage Program at Greystone Park Psychiatric Hospital, Greystone Park, New Jersey. In M. D. Farkas & W. A. Anthony (Eds.), *Psychiatric rehabilitation programs: Putting theory into practice* (pp. 57–69). Baltimore: Johns Hopkins University Press.

Creer, C., & Wing, J. K. (1974). *Schizophrenia at home.* London: Institute of Psychiatry.

Curran, T. (1980). A procedure for the assessment of social skills: The simulated social interaction test. In T. Curran & P. Monti (Eds.), *Social skills training: A practical handbook for assessment and treatment*. New York: Guilford Press.

Curry, J. (1981). A study in case management. *Community Support Service Journal, 2*, 15–17.

Cutler, D. L., Bloom, J. D., & Shore, J. H. (1981). Training psychiatrists to work with community support systems for chronically mentally ill persons. *American Journal of Psychiatry, 138*, 98–102.

Cutler, D. L., Tatum, E., & Shore, J. H. (1987). A comparison of schizophrenic patients in different community support treatment approaches. *Community Mental Health Journal, 23*, 103–113.

Davis, A. E., Dinitz, S., & Pasamanick, B. (1974). *Schizophrenics in the new custodial care community: Five years after the experiment.* Columbus: Ohio State University Press.

Davis, K. E. (1985). Presentation at state of the art training conference, July 11–12, Richmond, VA.

Deegan, P. E. (1988). Recovery: The lived experience of rehabilitation. *Psychosocial Rehabilitation Journal, 11*(4), 11–19.

Dellario, D. J. (1985). The relationship between mental health, vocational rehabilitation, interagency functioning, and outcome of psychiatrically disabled persons. *Rehabilitation Counseling Bulletin, 28*, 167–170.

Dellario, D. J., & Anthony, W. A. (1981). On the relative effectiveness of institutional and alternative placement for the psychiatrically disabled. *Journal of Social Issues, 37*(3), 21–33.

Dellario, D. J., Anthony, W. A., & Rogers, E. S. (1983). Client-practitioner agreement in the assessment of severely psychiatrically disabled persons' functional skills. *Rehabilitation Psychology, 28*, 243–248.

Dellario, D. J., Goldfield, E., Farkas, M. D., & Cohen, M. R. (1984). Functional assessment of psychiatrically disabled adults: Implications of research findings for functional skills training. In A. S. Halpern & M. J. Fuhrer (Eds.), *Functional assessment in rehabilitation* (pp. 239–252). Baltimore: Paul Brookes.

Dickey, B., Cannon, N. L., McGuire, T. G., & Gudeman, J. E. (1986). The quarterway house: A two-year cost study of an experimental residential program. *Hospital and Community Psychiatry, 37*, 1136–1143.

Dickey, B., & Goldman, H. H. (1986). Public health care for the chronically mentally ill: Financing operation costs: Issues and options for local leadership. *Administration in Mental Health, 14*, 63–77.

Dimsdale, J., Klerman, G., & Shershow, J. (1979). Conflict in treatment goals between patients and staff. *Social Psychiatry, 14*, 1–4.

Dincin, J. (1975). Psychiatric rehabilitation. *Schizophrenia Bulletin, 13*, 131–147.

Dincin, J. (1981). A community agency model. In J. A. Talbott (Ed.), *The chronic mentally ill* (pp. 212–226). New York: Human Sciences Press.

239

Dincin, J., & Witheridge, T. F. (1982). Psychiatric rehabilitation as a deterrent to recidivism. *Hospital and Community Psychiatry, 33*, 645–650.

Dion, G. L., & Anthony, W. A. (1987). Research in psychiatric rehabilitation: A review of experimental and quasi-experimental studies. *Rehabilitation Counseling Bulletin, 30*, 177–203.

Dion, G. L., & Cohen, M. R., Anthony, W. A., & Waternaux, C. S. (1988). Symptoms and functioning of patients with bipolar disorder six months after hospitalization. *Hospital and Community Psychiatry, 39*, 652–657.

Dion, G. L., & Dellario, D. J. (1988). Symptom subtypes in persons institutionalized with schizophrenia: Comparison of demographics, outcome and functional skills. *Rehabilitation Psychology, 33*, 95–104.

Dion, G. L., Dellario, D. J., & Farkas, M. D. (1982). The relationship of maintenance neuroleptic dosage levels to vocational functioning in severely psychiatrically disabled clients: Implications for rehabilitation practitioners. *Psychosocial Rehabilitation Journal, 6*(2), 29–35.

Distefano, M. K., & Pryer, M. W. (1970). Vocational evaluation and successful placement of psychiatric clients in a vocational rehabilitation program. *American Journal of Occupational Therapy, 24*, 205–207.

Docherty, J. P., Sims, S. G., & van Kammen, D. P. (1975). *Maintenance phenothiazine treatment in schizophrenia: A review*. Rockville, MD: National Institute of Mental Health.

Dodson, L. C., & Mullens, W. R. (1969). Some effects of jogging on psychiatric hospital patients. *American Corrective Therapy Journal, 23*, 130–134.

Doll, W. (1976). Family coping with the mentally ill: An unanticipated problem of deinstitutionalization. *Hospital and Community Psychiatry, 27*, 183–185.

Domergue, M. (1968). *Technical assistance: Theory, practice, and policies*. New York: Praeger.

Douzinas, N., & Carpenter, M. (1981). Predicting the community performance of vocational rehabilitation clients. *Hospital and Community Psychiatry, 32*, 309–412.

Dowell, D. A., & Ciarlo, J. A. (1983). Overview of the Community Mental Health Center's Program from an evaluation perspective. *Community Mental Health Journal, 19*, 95–128.

Dozier, M., Harris, M., & Bergman, H. C. (1987). Social network density and rehospitalization among young adult patients. *Hospital and Community Psychiatry, 38*, 61–65.

Eaton, L. F., & Menolascino, F. J. (1982). Psychiatric disorders in the mentally retarded: Types, problems, challenges. *American Journal of Psychiatry, 139*, 1297–1303.

Eisenberg, M. G., & Cole, H. W. (1986). A behavioral approach to job seeking for psychiatrically impaired persons. *Journal of Rehabilitation*, April/May/June, 46–49.

Ellsworth, R. B., Foster, L., Childers, B., Arthur, G., & Kroeker, D. (1968). Hospital and community adjustment as perceived by psychiatric patients, their families, and staff. *Journal of Consulting and Clinical Psychology, 32*, 1–41.

Englehardt, D. M., & Rosen, B. (1976). Implications of doing treatment for the social rehabilitation of schizophrenic patients. *Schizophrenia Bulletin, 2*, 454–462.

Erickson, R. C. (1975). Outcome studies in mental hospitals: A review. *Psychological Bulletin, 82*, 519–540.

Erickson, R. C., & Binder, L. M. (1986). Cognitive deficits among functionally psychotic patients: A rehabilitative perspective. *Journal of Clinical and Experimental Neuropsychology, 8*, 257–274.

Erickson, R. C., & Hyerstay, B. J. (1980). Historical perspectives on treatment of the mentally ill. In M. S. Gibbs, J. Lachermeyer, & J. Sigal (Eds.), *Community psychology: Theoretical and empirical approaches* (pp. 29–63). New York: Gardner Press.

Erlanger, H. S., & Roth, W. (1985). Disability policy. *American Behavioral Scientist, 28*, 319–346.

Ethridge, D. A. (1968). Pre-vocational assessment of the rehabilitation potential of psychiatric patients. *American Journal of Occupational Therapy, 22*, 161–167.

Evans, A. S., Bullard, D. M., & Solomon, M. H. (1961). The family as a potential resource in the rehabilitation of the chronic schizophrenic patient: A study of 60 patients and their families. *American Journal of Psychiatry, 117*, 1075–1082.

Fadden, G., Bebbington, P., & Kuipers, L. (1987). The burden of care: The impact of functional psychiatric illness on the patient's family. *British Journal of Psychiatry, 150*, 285–292.

Fairweather, G. W. (1971). *Methods of changing mental hospital programs.* (Progress Report to the National Institute of Mental Health No. R12–178887). East Lansing: Michigan State University.

Fairweather, G. W. (Ed.). (1980). *The Fairweather Lodge: A twenty-five year retrospective* (New Directions for Mental Health Service, No. 7). San Francisco: Jossey-Bass.

Falloon, I. R. H., Boyd, J. L., McGill, C. W., Strang, J. S., & Moss, H. B. (1982). Family management training in the community care of schizophrenia. In M. J. Goldstein (Ed.), *New developments in interventions with families of schizophrenics* (New Directions for Mental Health Services, No. 12, pp. 61- 77). San Francisco: Jossey-Bass.

Farkas, M. D., & Anthony, W. A. (1981). *The development of the rehabilitation model as a response to the shortcomings of the deinstitutionalization movement* (Monograph 1). Boston: Boston University, Center for Psychiatric Rehabilitation.

Farkas, M. D., & Anthony, W. A. (1987). Outcome analysis in psychiatric rehabilitation. In M. J. Fuhrer (Ed.), *Rehabilitation outcomes: Analysis and measurement* (pp. 43–56). Baltimore: Paul Brookes.

Farkas, M. D., & Anthony, W. A. (Eds.). (1989). *Psychiatric rehabilitation programs: Putting theory into practice.* Baltimore: Johns Hopkins University Press.

241

Farkas, M. D., Anthony, W. A., & Cohen, M. R. (1989). An overview of psychiatric rehabilitation: The approach and its programs. In M. D. Farkas & W. A. Anthony (Eds.), *Psychiatric rehabilitation programs: Putting theory into practice* (pp. 1–27). Baltimore: Johns Hopkins University Press.

Farkas, M. D., Cohen, M. R., & Nemec, P. B. (1988). Psychiatric rehabilitation programs: Putting concepts into practice. *Community Mental Health Journal, 24*, 7–21.

Farkas, M. D., O'Brien, W. F., & Nemec, P. B. (1988). A graduate level curriculum in psychiatric rehabilitation: Filling a need. *Psychosocial Rehabilitation Journal, 12*(2), 53–66.

Farkas, M. D., Rogers, E. S., & Thurer, S. (1987). Rehabilitation outcome of long-term hospital patients left behind by deinstitutionalization. *Hospital and Community Psychiatry, 38*, 864–870.

Farr, R. K. (1984). The Los Angeles Skid Row Mental Health Project. *Psychosocial Rehabilitation Journal, 8*(2), 64–76.

Felix, R. H. (1967). *Mental illness: Progress and prospect*. New York: Columbia University Press.

Fergus, E. O. (1980). Maintaining and advancing the lodge effort. In G. W. Fairweather (Ed.), *The Fairweather Lodge: A twenty-five year retrospective* (New Directions for Mental Health Service, No. 7, pp. 43–56). San Francisco: Jossey-Bass.

Field, G., Allness, D. J., & Knoedler, W. H. (1980). Application of the training in community living program to rural areas. *Journal of Community Psychology, 8*, 9–15.

Field, G., & Yegge, L. (1982). A client outcome study of a community support demonstration project. *Psychosocial Rehabilitation Journal, 6*(2), 15–22.

Fishbein, S. M. (1988). Partial care as a vehicle for rehabilitation of individuals with severe psychiatric disability. *Rehabilitation Psychology, 33*, 57–64.

Fishbein, S. M., & Cassidy, K. (1989). A system perspective on psychiatric rehabilitation: New Jersey. In M. D. Farkas & W. A. Anthony (Eds.), *Psychiatric rehabilitation programs: Putting theory into practice* (pp. 179-188). Baltimore: Johns Hopkins University Press.

Fisher, G., Landis, D., & Clark, K. (1988). Case management service provisions and client change. *Community Mental Health Journal, 24*, 134–142.

Fiske, D. W. (1983). The meta-analytic revolution in outcome research. *Journal of Consulting and Clinical Psychology, 51*, 65–70.

Foreyt, J. P., & Felton, G. S. (1970). Changes in behavior of hospitalized psychiatric patients in a milieu therapy setting. *Psychotherapy: Theory, Research and Practice, 7*, 139–141.

Forsyth, R. P., & Fairweather, G. W. (1961). Psychotherapeutic and other hospital treatment criteria: The dilemma. *Journal of Abnormal and Social Psychology, 62*, 598–604.

Fortune, J., & Eldredge, G. (1982). Predictive validation of the McCarron-Dial Evaluation System for psychiatrically disabled sheltered workshop workers. *Vocational Evaluation and Work Adjustment Bulletin, 15*, 136–141.

Fountain House. (1976). *Rehabilitation of the mental patient in the community.* Grant # 5T24MH14471. Rockville, MD: National Institute of Mental Health.

Fountain House (1985). *Evaluation of clubhouse model community based psychiatric rehabilitation: Final report for the National Institute of Handicapped Research* (Contract No. 300–84–0124). Washington, DC: National Institute of Handicapped Research.

Foy, D. W. (1984). Chronic alcoholism: Broad-spectrum clinical programming. In M. Mirabi (Ed.), *The chronically mentally ill: Research and services* (pp. 273–280). Jamaica, NY: Spectrum Publications.

Frank, J. D. (1981). Reply to Telch. *Journal of Consulting and Clinical Psychology, 49*, 476–477.

Franklin, J. L., Solovitz, B., Mason, M., Clemons, J. R., & Miller, G. E. (1987). An evaluation of case management. *American Journal of Public Health, 77*, 674–678.

Fraser, M. W., Fraser, M. E., & Delewski, C. H. (1985). The community treatment of the chronically mentally ill: An exploratory social network analysis. *Psychosocial Rehabilitation Journal, 9*(2), 35–41.

Freeman, H. E., & Simmons, O. G. (1963). *The mental patient comes home.* New York: John Wiley & Sons.

Frey, W. D. (1984). Functional assessment in the 80s: A conceptual enigma, a technical challenge. In A. S. Halpern & M. J. Fuhrer (Eds.), *Functional assessment in rehabilitation* (pp. 11–43). Baltimore: Paul Brookes.

Friday, J. C. (1987). *What is available in psychosocial rehabilitation training?* Atlanta: Southern Regional Education Board.

Gaebel, W., & Pietzcker, A. (1987). Prospective study of course of illness in schizophrenia: Part II: Prediction of outcome. *Schizophrenia Bulletin, 13*, 299–306.

Gaitz, L. M. (1984). Chronic mental illness in aged patients. In M. Mirabi (Ed.), *The chronically mentally ill: Research and services* (pp. 281–290). Jamaica, NY: Spectrum Publications.

Gardos, G., & Cole, J. O. (1976). Maintenance antipsychotic therapy: Is the cure worse than the disease? *American Journal of Psychiatry, 133*, 32–36.

Gay, R. D. (1983). The Georgia experience: Another perspective. *New Directions in Mental Health Services, 18*, 67–71.

Gelineau, U. A., & Evans, A. S. (1970). Volunteer case aides rehabilitate chronic patients. *Hospital and Community Psychiatry, 21*(3), 34–37.

George, L. K., Blazer, D. G., Hughes, D. C., & Fowler, N. (1989). Social support and the outcome of major depression. *British Journal of Psychiatry, 154*, 478–485.

Gerhart, U. C. (1985). Teaching social workers to work with the chronically mentally ill. In J. P. Bowker (Ed.), *Education for practice with the chronically mentally ill: What works?* Washington, DC: Council on Social Work Education.

Gittleman, M. (1974). Coordinating mental health systems. *American Journal of Public Health, 64,* 496–500.

Glaser, E. M., & Ross, U. L. (1971). *Increasing the utilization of applied research results* (NIMH Grant No. 5R12MHO925-2). Washington, DC: National Institute of Mental Health.

Glaser, E. M., & Taylor, S. (1969). *Factors influencing the success of applied research.* Final Report on Contract #43–67–1365, National Institute of Mental Health, Department of Health, Education & Welfare, Washington, DC.

Glasscote, R. M., Gudeman, J. E., & Elpers, R. (1971). *Halfway houses for the mentally ill: A study of programs and problems.* Washington, DC: Joint Information Service of the American Psychiatric Association and the National Association for Mental Health.

Goering, P. N., Farkas, M. D., Wasylenki, D. A., Lancee, W. J., & Ballantyne, R. (1988). Improved functioning for case management clients. *Psychosocial Rehabilitation Journal, 12*(1), 3–17.

Goering, P. N., Huddart, C., Wasylenki, D. A., & Ballantyne, R. (1989). The use of rehabilitation case management to develop necessary supports: Community Rehabilitation Services, Toronto, Ontario. In M. D. Farkas & W. A. Anthony (Eds.), *Psychiatric rehabilitation programs: Putting theory into practice* (pp. 197–207). Baltimore: Johns Hopkins University Press.

Goering, P. N., & Stylianos, S. K. (1988). Exploring the helping relationship between the schizophrenic client and rehabilitation therapist. *American Journal of Orthopsychiatry, 58*(2), 271–280.

Goering, P. N., Wasylenki, D. A., Farkas, M. D., Lancee, W. J., & Ballantyne, R. (1988). What difference does case management make? *Hospital and Community Psychiatry, 39,* 272–276.

Goering, P. N., Wasylenki, D. A., Lancee, W. J., & Freeman, S. J. J. (1984). From hospital to community: Six-month and two-year outcomes for 505 patients. *The Journal of Nervous and Mental Disease, 172,* 667–673.

Goffman, E. (1961). *Asylums: Essays on the social situation of mental patients and other inmates.* Garden City, NJ: Doubleday-Anchor.

Goin, M., Yamamoto, J., & Silverman, J. (1965). Therapy congruent with class linked expectations. *Archives of General Psychiatry, 133,* 455–470.

Goldberg, M. F., Evans, A. S., & Cole, K. H. (1973). The utilization and training of volunteers in a psychiatric setting. *British Journal of Social Work, 3*(1), 55–63.

Goldberg, S. C. (1980). Drug and psychosocial therapy in schizophrenia: Current status and research needs. *Schizophrenia Bulletin, 6,* 117–122.

Goldman, H. H., Burns, B. J., & Burke, J. D. (1980). Integrating primary health care and mental health services: a preliminary report. *Public Health Reports, 95,* 535–539.

Goldman, H. H., Gattozzi, A. A., & Taube, C. A. (1981). Defining and counting the chronically mentally ill. *Hospital and Community Psychiatry, 32*, 21–27.

Goldstein, A. P. (1981). *Psychological skill training*. New York: Pergamon Press.

Goldstein, A. P., & Kanfer, F. H. (Eds.). (1979). *Maximizing treatment gains: Transfer enhancement in psychotherapy*. New York: Academic Press.

Goldstein, M. J., & Kopeiken, H. S. (1981). Short- and long-term effects of combining drug and family therapy. In M. J. Goldstein (Ed.), *New developments in interventions with families of schizophrenics* (New Directions for Mental Health Services, No. 12, pp. 5–26). San Francisco: Jossey-Bass.

Goldstrom, I. D., & Manderscheid, R. W. (1982). The chronically mentally ill: A descriptive analysis from the Uniform Client Data Instrument. *Community Support Service Journal, 2*(3), 4–9.

Goldstrom, I. D., & Manderscheid, R. W. (1983). A descriptive analysis of community support program case managers serving the chronically mentally ill. *Community Mental Health Journal, 19*, 17–26.

Gomory, R. E. (1983). Technology development. *Science, 230*, 576–580.

Goodrick, P. (1988). *Strategies for state and local mental health system planning*. Washington, DC: COSMOS Corporation.

Gorin, S. S. (1986). Cost-outcome analysis and service planning in a CMHC. *Hospital and Community Psychiatry, 37*, 697–701.

Goss, A. M., & Pate, K. D. (1967). Predicting vocational rehabilitation success for psychiatric patients with psychological tests. *Psychological Reports, 21*, 725–730.

Green, H. J., Miskimins, R. W., & Keil, E. C. (1968). Selection of psychiatric patients for vocational rehabilitation. *Rehabilitation Counseling Bulletin, 11*, 297–302.

Greenblatt, M., Bererra, R. M., & Serafetinides, E. A. (1982). Social networks and mental health: An overview. *American Journal of Psychiatry, 139*, 977- 984.

Gregory, C. C., & Downie, M. N. (1968). Prognostic study of patients who left, returned, and stayed in a psychiatric hospital. *Journal of Counseling Psychology, 15*, 232–236.

Grella, C. E., & Grusky, O. (1989). Families of the seriously mentally ill and their satisfaction with services. *Hospital and Community Psychiatry, 40*, 831–835.

Griffiths, R. D. (1973). A standardized assessment of the work behavior of psychiatric patients. *British Journal of Psychiatry, 123*, 403–408.

Griffiths, R. (1974). Rehabilitation of chronic psychotic patients. *Psychological Medicine, 4*, 316–325.

Grinspoon, L., Ewalt, J. R., & Shader, R. I. (1972). *Schizophrenia: Pharmacotherapy and psychotherapy*. Baltimore: Williams and Wilkins.

245

Grob, S. (1970). Psychiatric social clubs come of age. *Mental Hygiene, 54,* 129–136.

Grob, S. (1983). Psychosocial rehabilitation centers: Old wine in a new bottle. In I. Barofsky & R. D. Budson (Eds.), *The chronic psychiatric patient in the community: Principles of treatment* (pp. 265–280). Jamaica, NY: Spectrum Publications.

Growick, B. (1979). Another look at the relationship between vocational and nonvocational client change. *Rehabilitation Counseling Bulletin, 23,* 136-139.

Grusky, O., & Tierney, K. (1989). Evaluating the effectiveness of countywide mental health care systems. *Community Mental Health Journal, 25,* 3–19.

Grusky, O., Tierney, K., Anspach, R., Dans, D., Holstein, J., Unruh, D., & Vandewater, S. (1987). Descriptive evaluation of community support programs. *International Journal of Mental Health, 15*(4), 26–43.

Grusky, O., Tierney, K., Holstein, J., Anspach, R., Dans, D., Unruh, D., Webster, S., Vandewater, S., & Allen, H. (1985). Models of local mental health delivery systems. *American Behavioral Scientist, 28*(5), 685–703.

Gurel, L., & Lorei, T. W. (1972). Hospital and community ratings of psychopathology as predictors of employment and readmission. *Journal of Counseling and Clinical Psychology, 34,* 286–291.

Hall, J. D., Smith, K., & Shimkunas, A. (1966). Employment problems of schizophrenic patients. *American Journal of Psychiatry, 123,* 536–540.

Hamilton, L. S., & Muthard, J. E. (1975). *Research utilization specialists in vocational rehabilitation* (Monograph). Gainesville, FL: Rehabilitation Research Institute.

Hammaker, R. (1983). A client outcome evaluation of the statewide implementation of community support services. *Psychosocial Rehabilitation Journal, 7*(1), 2–10.

Harding, C. M., Brooks, G. W., Ashikaga, T., Strauss, J. S., & Breier, A. (1987a). The Vermont longitudinal study of persons with severe mental illness: I. Methodology, study sample, and overall status 32 years later. *American Journal of Psychiatry, 144,* 718–726.

Harding, C. M., Brooks, G. W., Ashikaga, T., Strauss, J. S., & Breier, A. (1987b). The Vermont longitudinal study of persons with severe mental illness: II. Long-term outcome of subjects who retrospectively met *DSM-III* criteria for schizophrenia. *American Journal of Psychiatry, 144,* 727–735.

Harding, C. M., Strauss, J. S., Hafez, H., & Lieberman, P. B. (1987). Work and mental illness: I. Toward an integration of the rehabilitation process. *The Journal of Nervous and Mental Disease, 175,* 317–326.

Harding, C. M., Zubin, J., & Strauss, J. S. (1987). Chronicity in schizophrenia: Fact, partial fact, or artifact? *Hospital and Community Psychiatry, 38,* 477–486.

Harrand, G. (1967). Rehabilitation program for chronic patients: Testing the potential for independence. *Hospital and Community Psychiatry, 18,* 376–377.

Harris, M., & Bergman, H. C. (1985). Networking with young adult chronic patients. *Psychosocial Rehabilitation Journal, 8*(3), 28–35.

Harris, M., & Bergman, H. C. (1987a). Case management with the chronically mentally ill: A clinical perspective. *American Journal of Orthopsychiatry, 57,* 296–302.

Harris, M., & Bergman, H. C. (1987b). Differential treatment planning for young adult chronic patients. *Hospital and Community Psychiatry, 38,* 638–643.

Harris, M., & Bergman, H. C. (1988a). Capitation financing for the chronic mentally ill: A case management approach. *Hospital and Community Psychology, 39,* 68–72.

Harris, M., & Bergman, H. C. (1988b). Clinical case management for the chronically mentally ill: A conceptual analysis. In M. Harris & L. L. Bachrach (Eds.), *Clinical case management* (New Directions for Mental Health Services, No. 40, pp. 5–13). San Francisco: Jossey-Bass.

Harris, M., & Bergman, H. C. (1988c). Misconceptions about the use of case management services by the chronic mentally ill: A utilization analysis. *Hospital and Community Psychology, 39,* 1276–1280.

Harris, M., Bergman, H. C., & Bachrach, L. L. (1987). Individualized network planning for chronic psychiatric patients. *Psychiatry Quarterly, 58*(1), 51-56.

Hatfield, A. B. (1978). Psychological costs of schizophrenia to the family. *Social Work, 23,* 355–359.

Hatfield, A. B. (1979). The family as partner in the treatment of mental illness. *Hospital and Community Psychiatry, 30,* 338–340.

Hatfield, A. B. (1981). Self-help groups for families of the mentally ill. *Social Work, 26,* 408–413.

Hatfield, A. B. (1983). What families want of family therapists. In W. McFarlane (Ed.), *Family therapy in schizophrenia.* New York: Guilford.

Hatfield, A. B., Fierstein, R., & Johnson, D. M. (1982). Meeting the needs of families of the psychiatrically disabled. *Psychosocial Rehabilitation Journal, 6*(1), 27–40.

Hatfield, A. B., Spaniol, L. J., & Zipple, A. M. (1987). Expressed emotion: A family perspective. *Schizophrenia Bulletin, 13,* 221–226.

Havari, D. (1974). *The role of the technical assistance expert.* Organization for Economic Cooperation and Development, Paris.

Havelock, R. G. (1971). *Planning for innovation through dissemination and utilization of knowledge.* Ann Arbor: University of Michigan, Institute for Social Research.

Havelock, R. G., & Benne, K. D. (1969). An exploratory study of knowledge utilization. In W. G. Bennis, K. D. Benne, & R. Chien (Eds.), *The planning of change* (2nd ed.). New York: Holt, Rinehart & Winston.

247

Havens, L. L. (1967). Dependence: Definitions and strategies. *Rehabilitation Record*, March/April, 23–28.

Heap, F. R., Boblitt, E. W., Moore, H., & Hord, E. J. (1970). Behavior milieu therapy with chronic neuropsychiatric patients. *Journal of Abnormal Psychology, 76*, 349–354.

Hersen, M. (1979). Limitations and problems in the clinical applications of behavioral techniques in psychiatric settings. *Behavioral Therapy, 10*, 65-80.

Hersen, M., & Bellack, A. S. (1976). Social skills training for chronic psychiatric patients: Rationale, research, findings and further directions. *Comprehensive Psychiatry, 17*, 559–580.

Hersen, M., & Bellack, A. (1977). The assessment of social skills. In A. Ciminero, K. Calhoun, & H. Adams (Eds.), *Handbook of behavioral assessment* (pp. 509–554). New York: John Wiley & Sons.

Herz, M. I., Spitzer, R. L., Gibbon, M., Greerspan, K., & Reibel, S. (1974). Individual versus group aftercare treatment. *American Journal of Psychiatry, 303*, 808–812.

Herz, M. I., Szymanski, H. V., Simon, J. C. (1982). Intermittent medication for stable schizophrenic outpatients: An alternative to maintenance medication. *American Journal of Psychiatry, 139*, 918–922.

Hibler, M. (1978). The problems as seen by the patient's family. *Hospital and Community Psychiatry, 29*(1), 32–33.

Hillhouse-Jones, L. (1984, August). Psychiatric rehabilitation training: A trainee's perspective. *Florida Community Support Network Newsletter, 1*, 8.

Hoffman, D. A. (1980). *The differential effects of self-monitoring, self-reinforcement and performance standards on the production output, job satisfaction and attendance of vocational rehabilitation clients.* Unpublished doctoral dissertation, Catholic University of America, Washington, DC.

Hogarty, G. E., McEvoy, J. P., Munetz, M., Di Barry, A. L., Bartone, P., Cather, R., Cooley, S. J., Ulrich, R. F., Carter, M., & Madonia, M. J. (1988). Dose of fluphenazine, familial expressed emotion, and outcome in schizophrenia: Results of a two-year controlled study. *Archives of General Psychiatry, 45*, 797–805.

Hogarty, G. E., Anderson, C. M., Reiss, D. J., Kornblith, S. J., Greenwald, D. P., Javna, C. D., & Madonia, M. J. (1986). Family psychoeducation, social skills training, and maintenance chemotherapy in the aftercare treatment of schizophrenia: I. One-year effects of a controlled study on relapse and expressed emotion. *Archives of General Psychiatry, 43*, 633–642.

Holcomb, W. R., & Ahr, P. R. (1986). Clinician's assessments of the service needs of young adult patients in public mental health care. *Hospital and Community Psychiatry, 37*, 908–913.

Hollingsworth, R., & Foreyt, J. (1975). Community adjustment of released token economy patients. *Journal of Behavior Therapy and Experimental Psychiatry, 6*, 271–274.

Holroyd, J., & Goldenberg, I. (1978). The use of goal attainment scaling to evaluate a ward-treatment program for disturbed children. *Journal of Clinical Psychology, 34*, 732–739.

Hoult, J. (1986). Community care of the acutely mentally ill. *British Journal of Psychiatry, 149,* 137–144.

Hutchinson, D. S., Kohn, L., & Unger, K. V. (1989). A university-based psychiatric rehabilitation program for young adults: Boston University. In M. D. Farkas & W. A. Anthony (Eds.), *Psychiatric rehabilitation programs: Putting theory into practice* (pp. 147–157). Baltimore: Johns Hopkins University Press.

Intagliata, J. (1982). Improving the quality of care for the chronically mentally disabled: The role of case management. *Schizophrenia Bulletin, 8,* 655–674.

Intagliata, J., & Baker, F. (1983). Factors affecting case management services for the chronically mentally ill. *Administration in Mental Health, 11,* 75-91.

Ivey, A. E. (1973). Media therapy: Education change planning for psychiatric patients. *Journal of Counseling Psychology, 20,* 338–343.

Jacobs, H. E., Kardashian, S., Kreinbring, R. K., Ponder, R., & Simpson, A. R. (1984). A skills oriented model for facilitating employment among psychiatrically disabled persons. *Rehabilitation Counseling Bulletin, 28,* 87–96,

Jacobs, M., & Trick, O. (1974). Successful psychiatric rehabilitation using an inpatient teaching laboratory: A one-year follow-up study. *American Journal of Psychiatry, 131,* 145–148.

Jeger, A. M., & McClure, G. (1980). The effects of a behavioral training program on nonprofessional endorsement of the "psychosocial" model. *Journal of Community Psychology, 8,* 49–53.

Jensen, K., Spangaard, P., Juel-Neilsen, N., & Voag, V. H. (1978). Experimental psychiatric rehabilitation unity. *International Journal of Social Psychiatry, 24,* 53–57.

Jerrell, J. M., & Larsen, J. K. (1985). How community mental health centers deal with cutbacks and competition. *Hospital and Community Psychiatry, 36,* 1169–1174.

Joint Commission on Accreditation of Hospitals. (1976). *Accreditation of community mental health service programs.* Chicago: Author.

Jung, H. F., & Spaniol, L. J. (1981). *Planning the utilization of new knowledge and skills: Some basic principles for researchers, administrators and practitioners.* Unpublished manuscript, Boston University, Center for Psychiatric Rehabilitation, Boston.

Kahn, R. L., & Quinn, R. P. (1977). *Mental health, social adjustment, and metropolitan problems.* Research proposal, University of Michigan, Ann Arbor.

Kane, J. M. (1987). Low-dose and intermittent neuroleptic treatment strategies for schizophrenia: An interview with John Kane. *Psychiatric Annals, 17,* 125–130.

Katkin, S., Ginsburg, M., Rifkin, J. J., & Scott, J. T. (1971). Effectiveness of female volunteers in the treatment of out-patients. *Journal of Counseling Psychology, 18*, 97–100.

Katkin, S., Zimmerman, V., Rosenthal, J., & Ginsburg, M. (1975). Using volunteer therapists to reduce hospital readmissions. *Hospital and Community Psychiatry, 26*, 151–153.

Katz-Garris, L., McCue, M., Garris, R. P., & Herring, J. (1983). Psychiatric rehabilitation: An outcome study. *Rehabilitation Counseling Bulletin, 26*, 329–335.

Keith, S. J., & Matthews, S. M. (1984). Research overview. In J. A. Talbott (Ed.), *The chronic mental patient: Five years later* (pp. 7–13). Orlando, FL: Grune & Stratton.

Kelner, F. B. (1984). A rehabilitation approach to program diagnosis in technical assistance consultation. *Psychosocial Rehabilitation Journal, 7*(3), 32–43.

Kennedy, E. M. (1989). *Community based care for the mentally ill: Simple justice*. Unpublished manuscript, Boston University, Center for Psychiatric Rehabilitation, Boston.

Kerlinger, F. M. (1964). *Foundations of behavioral research*. New York: Holt, Rinehart & Winston.

Kerr, N., & Meyerson, L. (1987). Independence as a goal and a value of people with physical disabilities: Some caveats. *Rehabilitation Psychology, 12*, 173–180.

Kiesler, C. A. (1982). Mental hospitals and alternative care. *American Psychologist, 37*, 349–360.

Killilea, M. (1976). Mutual help organizations: Interpretations in the literature. In G. Kapplan & M. Killilea (Eds.), *Support systems and mutual help*. New York: Grune & Stratton.

Killilea, M. (1982). Interaction of crisis theory, coping strategies, and social support systems. In H. C. Schulberg & M. Killilea (Eds.), *The modern practice of community mental health*. San Francisco: Jossey-Bass.

Kline, M. N., & Hoisington, V. (1981). Placing the psychiatrically disabled: A look at work values. *Rehabilitation Counseling Bulletin, 24*, 366–369.

Koumans, A. J. (1969). Reaching the unmotivated patient. *Mental Hygiene, 53*(2), 298–300.

Kunce, J. T. (1970). Is work therapy really therapeutic? *Rehabilitation Literature, 31*, 297–299.

Kurtz, L., Bagarozzi, D., & Pollane, L. (1984). Case management in mental health. *Health and Social Work, 9*, 201–211.

Lamb, H. R. (1982). *Treating the long term mentally ill*. San Francisco: Jossey-Bass.

Lamb, H. R., & Oliphant, E. (1979). Parents of schizophrenics: Advocates for the mentally ill. In L. I. Stein (Ed.), *Community support systems for the long-term patient* (New Directions for Mental Health Services, No. 2, pp. 85- 92). San Francisco: Jossey-Bass.

Lang, E., & Rio, J. (1989). A psychiatric rehabilitation vocational program in a private psychiatric hospital: The New York Hospital-Cornell Medical Center, Westchester Division, White Plains, NY. In M. D. Farkas & W. A. Anthony (Eds.), *Psychiatric rehabilitation programs: Putting theory into practice* (pp. 86–98). Baltimore: Johns Hopkins University Press.

Langsley, D. G., & Kaplan, D. M. (1968). *The treatment of families in crisis.* New York: Grune & Stratton.

Langsley, D. G., Machotka, P., & Flomenhaft, K. (1971). Avoiding mental hospital admission: A follow-up study. *American Journal of Psychiatry, 129*, 1391–1394.

Lannon, P. B., Banks, S. M., & Morrissey, J. P. (1988). Community tenure patterns of the New York State CSS population: A longitudinal impact assessment. *Psychosocial Rehabilitation Journal, 11*(4), 47–60.

LaPaglia, J. E. (1981). *The use of role-play strategies to teach vocationally related social skills to mentally handicapped persons: Three studies of training and generalization.* Unpublished doctoral dissertation, Vanderbilt University, Nashville, TN.

Larsen, J. K. (1987). Community mental health services in transition. *Community Mental Health Journal, 23*, 16–25.

Lazare, A., Eisenthal, S., & Wasserman, L. (1975). The customer approach to patienthood: Attending to patient requests in a walk-in clinic. *Archives of General Psychiatry, 32*, 553–558.

Lecklitner, G. L., & Greenberg, P. D. (1983). Promoting the rights of the chronically mentally ill in the community: A report on the Patient Rights Policy Research Project. *Mental Disability Law Reporter, 7*, 422–430.

Leff, J., Kuipers, L., Berkowitz, R., Eberbein-Vries, R., & Sturgeon, D. A. (1982). Controlled trial of social intervention in the families of schizophrenic patients. *British Journal of Psychiatry, 141*, 121–134.

Lefley, H. P. (1987). Aging parents as caregivers of mentally ill adult children: An emerging social problem. *Hospital and Community Psychiatry, 38*, 1063–1069.

Lehman, A. F. (1987). Capitation payment and mental health care: A review of the opportunities and risks. *Hospital and Community Psychiatry, 38*, 31–38.

Leitner, L. & Drasgow, J. (1972). Battling recidivism. *Journal of Rehabilitation*, July/August, 29–31.

Levine, I. S., & Fleming, M. (1984). *Human resource development: Issues in case management.* Rockville, MD: National Institute of Mental Health.

Leviton, G. (1973). Professional and client viewpoints on rehabilitation issues. *Rehabilitation Psychology, 20*, 1–80.

Lewington, J. (1975). Volunteer case aides in the U.S.A. *International Journal of Social Psychiatry, 21*(3), 205–213.

Liberman, R. P., & Foy, D. W. (1983). Psychiatric rehabilitation for chronic mental patients. *Psychiatric Annals*, *13*, 539–545.

Liberman, R. P., Mueser, K. T., & Wallace, C. J. (1986). Social skills training for schizophrenic individuals at risk for relapse. *American Journal of Psychiatry*, *143*, 523–526.

Liberman, R. P., Mueser, K. T., Wallace, C. J., Jacobs, H. E., Eckman, T., & Massel, H. K. (1986). Training skills in the psychiatrically disabled: Learning coping and competence. *Schizophrenia Bulletin*, *12*, 631–647.

Lieberman, M. A. (1986). Social supports: The consequences of psychologizing: A commentary. *Journal of Consulting and Clinical Psychology*, *54*, 461–465.

Linn, M. W., Caffey, E. M., Klett, J., Hogarty, G. E., & Lamb, H. R. (1979). Day treatment and psychotropic drugs in the aftercare of schizophrenic patients. *Archives of General Psychiatry*, *36*, 1055–1066.

Locke, E. A., Shaw, K. N., & Saari, L. M., & Latham, G. P. (1981). Goal setting and task performance: 1969–1980. *Psychological Bulletin*, *90*, 125–152.

Lorei, T. W. (1967). Prediction of community stay and employment for released psychiatric patients. *Journal of Consulting Psychology*, *31*, 349–357.

Lorei, T. W., & Gurel, L. (1973). Demographic characteristics as predictors of post-hospital employment and readmission. *Journal of Consulting and Clinical Psychology*, *40*, 426–430.

Makas, E. (1980). Increasing counselor-client communication. *Rehabilitation Literature*, September/October, 235–238.

Marlowe, H. A., Marlowe, J. L., & Willets, R. (1983). The mental health counselor as a case manager: Implications for working with the chronically mentally ill. *American Mental Health Counselors Association Journal*, *5*, 184–191.

Marlowe, H. A., Spector, P. E., & Bedell, J. R. (1983). Implementing a psychosocial rehabilitation program in a state mental hospital: A case study of organization change. *Psychosocial Rehabilitation Journal*, *6*(3), 2–11.

Marlowe, H. A., & Weinberg, R. (1983). (Eds.). *Proceedings of the 1982 CSP Region 4 conference*. Tampa: University of South Florida.

Marshall, C. (1989). Skill teaching as training in rehabilitation counselor education. *Rehabilitation Education*, *3*, 19–26.

Martin, H. R. (1959). A philosophy of rehabilitation. In M. Greenblatt & B. Simon (Eds.), *Rehabilitation of the mentally ill*. Washington, DC: American Association for the Advancement of Science.

Matthews, S. M., Roper, M. T., Mosher, L. R., & Menn, A. Z. (1979). A non-neuroleptic treatment for schizophrenia: Analysis of the two-year post-discharge risk of relapse. *Schizophrenia Bulletin*, *5*, 322–332.

Matthews, W. C. (1979). Effects of a work activity program on the self-concept of chronic schizophrenics. *Dissertations Abstracts International*, *41*, 358B. (University Microfilms No. 8816281, 98).

McClure, D. P. (1972). Placement through improvement of client's job-seeking skills. *Journal of Applied Rehabilitation Counseling*, *3*, 188–196.

McCreadie, R. G., & Phillips, K. (1988). The Nittsdale Schizophrenic Survey VII. Does relatives' high expressed emotion predict relapse? *British Journal of Psychiatry, 152*, 477–481.

McCue, M., Katz-Garris, L. (1985). A survey of psychiatric rehabilitation counseling training needs. *Rehabilitation Counseling Bulletin, 28*, 291–297.

McGlashan, T. H. (1987). Recovery style from mental illness and long-term outcome. *The Journal of Nervous and Mental Disease, 175*, 681–685.

McNees, M. P., Hannah, J. T., Schnelle, J. F., & Bratton, K. M. (1977). The effects of aftercare programs on institutional recidividsm. *Journal of Community Psychology, 5*, 128–133.

Mental Health Policy Resource Center. (1988). A typology for mental health case management for persons with severe mental illness. In *Report on the state-of-the-art of case management programs*. Washington, DC: Author.

Meyerson, A., & Herman, G. (1983). What's new in aftercare? A review of recent literature. *Hospital and Community Psychiatry, 34*, 333–342.

Michaux, M. H., Chelst, M. R., Foster, S. A., & Pruin, R. (1972). Day and full-time psychiatric treatment: A controlled comparison. *Current Therapy Research, 14*, 279–292.

Miles, D. G. (1983). The Georgia experience: Unifying state and local services around the Balanced Service System Model. In J. A. Talbott (Ed.), *Unified mental health systems: Utopia unrealized* (New Directions for Mental Health Services, No. 18, pp. 53–65). San Francisco: Jossey-Bass.

Miles, P. G. (1967). A research-based approach to psychiatric rehabilitation. In L. M. Roberts (Ed.), *The role of vocational rehabilitation in community mental health*. Washington, DC: Rehabilitation Services Administration.

Miller, S., & Wilson, N. (1981). The case for performance contracting. *Administration in Mental Health, 8*, 185–193.

Miller, T. W. (1981). A model for training schizophrenics and families to communicate more effectively. *Hospital and Community Psychiatry, 32*, 870-871.

Minkoff, K. (1979). A map of chronic patients. In J. Talbott (Ed.), *The chronic mental patient*. Washington, DC: American Psychiatric Association.

Minkoff, K. (1987). Resistance of mental health professionals to working with the chronic mentally ill. In A. T. Meyerson (Ed.), *Barriers to treating the chronic mentally ill* (New Directions for Mental Health Services, No. 33, pp. 3–20). San Francisco: Jossey-Bass.

Mintz, L. I., Liberman, R. P., Miklowitz, D. J., & Minty, J. (1987). Expressed emotion: A call for partnership among relatives, patients, and professionals. *Schizophrenia Bulletin, 13*, 227–235.

Miskimins, R., Wilson, T., Berry, K., Oetting, E., & Cole, C. (1969). Person-placement congruence: A framework for vocational counselors. *Personnel and Guidance Journal*, April, 789–793.

253

Mitchell, J. E., Pyle, R. L., & Hatsukami, D. (1983). A comparative analysis of psychiatric problems listed by patients and physicians. *Hospital and Community Psychiatry, 34*(9), 848–849.

Mitchell, R. E. (1982). Social networks and psychiatric clients: The personal and environmental context. *American Journal of Community Psychology, 4*, 387–401.

Modrcin, M., Rapp, C. A., & Chamberlain, R. (1985). *Case management and psychiatrically disabled individuals: Curriculum and training program.* Lawrence: University of Kansas School of Social Welfare.

Modrcin, M., Rapp, C. A., Poertner, J. (1988). The evaluation of case management services with the chronically mentally ill. *Evaluation and Program Planning, 11*, 307–314.

Möller, H., von Zerssen, D., Werner-Eilert, K., & Wuschenr-Stockheim, M. (1982). Outcome in schizophrenic and similar paranoid psychoses. *Schizophrenic Bulletin, 8*, 99–108.

Monti, P. M., & Fingeret, A. L. (1987). Social perception and communication skills among schizophrenics and nonschizophrenics. *Journal of Clinical Psychology, 43*, 197–205.

Monti, P. M., Fink, E., Norman, W., Curran, J. P., Hayes, S., & Caldwell, A. (1979). The effects of social skills training groups and social skills bibliotherapy with psychiatric patients. *Journal of Consulting and Clinical Psychology, 47*, 189–191.

Moore, D. J., Davis, M., & Mellon, J. (1985). *Academia's response to state mental health system needs.* Boulder, CO: Western Interstate Commission for Higher Education.

Morin, R. C., & Seidman, E. (1986). A social network approach and the revolving door patient. *Schizophrenia Bulletin, 12*, 262–273.

Morrison, R. L., & Bellack, A. S. (1984). Social skills training. In A. S. Bellack (Ed.), *Schizophrenia: Treatment, management, and rehabilitation* (pp. 247–279). Orlando, FL: Grune & Stratton.

Morrissey, J. P., Tausig, M., & Lindsey, M. L. (1985). Community mental health delivery systems. *American Behavioral Scientist, 28*(5), 704–720.

Mosher, L. R. (1983). Alternatives to psychiatric hospitalization: Why has research failed to be translated into practice? *New England Journal of Medicine, 309*(25), 1579–1580.

Mosher, L. R. (1986). The current status of the community support program: A personal assessment. *Psychosocial Rehabilitation Journal, 9*(3), 3–14.

Mosher, L. R., & Keith, S. J. (1979). Research on the psychosocial treatment of schizophrenia: Summary report. *American Journal of Psychiatry, 136*, 623-631.

Mosher, L. R., & Menn, A. Z. (1978). Community residential treatment for schizophrenia: Two-year follow-up. *Hospital and Community Psychiatry, 29*, 715–723.

254

Mowbray, C. T., & Freddolino, P. (1986). Consulting to implement nontraditional community programs for the long-term mentally disabled. *Administration in Mental Health, 14*, 122–134.

Mulkern, V. M., & Manderscheid, R. W. (1989). Characteristics of community support program clients in 1980 and 1984. *Hospital and Community Psychiatry, 40*, 165–172.

Muller, J. B. (1981). Alabama community support project evaluation of the implementation and initial outcome of a model case manager system. *Community Support Service Journal, 6*, 1–4.

Muthard, J. E. (1980). *Putting rehabilitation knowledge to use.* (Rehabilitation Monograph Number 11.) Gainesville, FL: Rehabilitation Research Institute.

Mynks, D. A., & Graham, R. S. (1989). Starting a new psychiatric rehabilitation residential program: ReVisions, Inc., Catonsville, Maryland. In M. D. Farkas & W. A. Anthony (Eds.), *Psychiatric rehabilitation programs: Putting theory into practice* (pp. 47–57). Baltimore: Johns Hopkins University Press.

Nadler, D. A. (1977). How information changes behavior. *Feedback and organization development using data based methods.* Reading, MA: Addison-Wesley.

National Institute of Handicapped Research (1980). A skills training approach in psychiatric rehabilitation. *Rehabilitation Research Brief, 4*(1). Washington, DC.

National Institute of Mental Health. (1980). *Announcement of community support system strategy development and implementation grants.* Rockville, MD: Author.

National Institute of Mental Health. (1987). *Toward a model plan for a comprehensive, community-based mental health system.* Rockville, MD: Division of Education and Service Systems Liaison.

Nemec, P. B. (1983). *Technical assistance.* Unpublished manuscript, Boston University, Center for Psychiatric Rehabilitation, Boston.

Nemec, P. B., Forbess, R., Cohen, M. R., Farkas, M. D., Rogers, E. S., & Anthony, W. A. (1990). *Technical assistance to psychiatric rehabilitation programs.* Manuscript submitted for publication.

Nemec, P. B., & Furlong-Norman, K. (1989). Supports for psychiatrically disabled persons. In M. D. Farkas & W. A. Anthony (Eds.), *Psychiatric rehabilitation programs: Putting theory into practice* (pp. 192–197, 223-225). Baltimore: Johns Hopkins University Press.

New Jersey Division of Mental Health and Hospitals. (1980). *Rules and regulations governing community mental health services and state aid.*

New York State Office of Mental Health. (1979). *CSS-100. Community support systems, NIMH client assessment.* Unpublished manuscript, Albany.

Öhman, A., Nordby, H., & d'Elia, G. (1986). Orienting and schizophrenia: Stimulus significance, attention, and distraction in a signaled reaction time task. *Journal of Abnormal Psychology, 95*, 326–334.

Parker, G., Johnston, P., & Hayward, L. (1988). Parental "expressed emotion" as a predictor of schizophrenic relapse. *Archives of General Psychiatry*, *45*, 806–813.

Pasamanick, B., Scarpitti, F., & Dinitz, S. (1967). *Schizophrenics in the community*. New York: Appleton-Century-Crofts.

Patterson, R., & Teigen, J. (1973). Conditional and post-hospital generalization of non-delusional responses in chronic psychotic patients. *Journal of Applied Behavior Analysis*, *6*, 65–70.

Paul, G. L. (1984). Residential treatment programs and aftercare for the chronically institutionalized. In M. Mirabi (Ed.), *The chronically mentally ill: Research and services* (pp. 239–269). Jamaica, NY: Spectrum Publications.

Paul, G. L., & Lentz, R. J. (1977). *Psychosocial treatment of chronic mental patients: Milieu versus social-learning programs*. Cambridge, MA: Harvard University Press.

Paul, G. L., Tobias, L. L., & Holly, B. L. (1972). Maintenance psychotropic drugs in the presence of active treatment programs. *Archives of General Psychiatry*, *27*, 106–115.

Pelletier, J. R., Rogers, E. S., & Thurer, S. (1985). The mental health needs of individuals with severe physical disability: A consumer advocate perspective. *Rehabilitation Literature*, *46*, 186–193.

Pelz, D. C., & Munson, R. C. (1980, January). *A framework for organizational innovating*. Unpublished manuscript, University of Michigan, Ann Arbor.

Pepper, B., & Ryglewicz, H. (1982). The young adult chronic patient. In B. Pepper & H. Ryglewicz (Eds.), *The young adult chronic patient* (New Directions for Mental Health Services, No. 14, pp. 121–124). San Francisco: Jossey-Bass.

Pepper, B., & Ryglewicz, H. (1983). Unified services: A New York state perspective. In J. A. Talbott (Ed.), *Unified mental health systems: Utopia unrealized* (New Directions for Mental Health Services, No. 18, pp. 39–47). San Francisco: Jossey-Bass.

Pepper, B., & Ryglewicz, H. (Eds.). (1984). *Advances in treating the young adult chronic patient* (New Directions for Mental Health Services, No. 21). San Francisco: Jossey-Bass.

Pepper, B., & Ryglewicz, H. (1988). Taking issue: What's in a diagnosis— and what isn't. *Hospital and Community Psychiatry*, *39*, 7.

Peters, B. (1985, Fall). Labels. *The Disability Rag*, Fall, 33.

Peterson, R. (1979). What are the needs of the chronic mental patient? In J. A. Talbott (Ed.), *The chronic mental patient: Problems, solutions, and recommendation for a public policy*. Washington, DC: American Psychiatric Press.

Pierce, J., & Blanch, A. K. (1989). A statewide psychosocial rehabilitation system: Vermont. In M. D. Farkas and W. Anthony (Eds.), *Psychiatric rehabilitation programs: Putting theory into practice* (pp. 170–179). Baltimore: Johns Hopkins University Press.

Pierce, R. M., & Drasgow, J. (1969). Teaching facilitative interpersonal functioning to psychiatric patients. *Journal of Counseling Psychology*, *16*, 295–298.

Pietzcker, A., & Gaebel, W. (1987). Prospective study of course of illness in schizophrenia: Part I. Outcome at 1 year. *Schizophrenia Bulletin, 13*, 287-297.

Pincus, H. A. (1980). Linking general health and mental health systems of care: Conceptual model of implementation. *American Journal of Psychiatry, 137*, 315–320.

Polak, P. R. (1978). A comprehensive system of alternatives to psychiatric hospitalization. In L. I. Stein & M. A. Test (Eds.), *Alternatives to mental hospital treatment*. New York: Plenum Press.

Power, C. (1979). The time-sample behavior checklist: Observational assessment of patient functioning. *Journal of Behavior Assessment, 1*(3), 199–210.

Power, P. W., & Dell Orto, A. E. (Eds.). (1980). *The role of the family in the rehabilitation of the physically disabled*. Austin, TX: PRO-ED.

Prazak, J. A. (1969). Learning job seeking interview skills. In J. Krumboltz & C. Thoreson (Eds.), *Behavioral Counseling* (pp. 414–428). New York: Rinehart & Winston.

Propst, R. N. (1985). The Fountain House national training program. *Community Support Network News 2*(2), 2.

Rapp, C. A. (1985). Research on the chronically mentally ill: Curriculum implications. In J. P. Bowker (Ed.), *Education for practice with the chronically mentally ill: What works?* (pp. 19–49). Washington, DC: Council on Social Work Education.

Rapp, C. A., & Chamberlain, R. (1985). Case management services for the chronically mentally ill. *Social Work, 26*, 417–422.

Rapp, C. A., & Wintersteen, R. T. (1989). The strengths model of case management: Results from twelve demonstrations. *Psychosocial Rehabilitation Journal, 13*(1), 23–32.

Rappaport, J., Seidman, E., Toro, P. A., McFadden, L. S., Reischl, T. M., Roberts, L. J., Salem, D. A., Stein, C. H., & Zimmerman, M. A. (1985, Winter). Collaborative research with a mutual help organization. *Social Policy*, 12–24.

Redfield, J. (1979). Clinical frequencies recording systems: Standardizing staff observations by event recording. *Journal of Behavior Assessment, 1*(3), 199–210.

Reinke, B., & Greenley, J. R. (1986). Organizational analysis of three community support program models. *Hospital and Community Psychiatry, 37*, 624–629.

Reischl, T. M., & Rappaport, J. (1988, August). *Participation in mutual help groups and coping with acute stressors*. Paper presented at the Annual Meeting of the American Psychological Association.

Reiss, S. (1987). Symposium overview: Mental health and mental retardation. *Mental Retardation, 25*, 323–324.

Retchless, M. H. (1967). Rehabilitation programs for chronic patients: Stepping stones to the community. *Hospital and Community Psychiatry, 18*, 377–378.

Rice, D. H., Seibold, M., & Taylor, J. (1989). Psychiatric rehabilitation in a residential setting: Alternatives Unlimited, Inc., Whitinsville, MA. In M. D. Farkas & W. A. Anthony (Eds.), *Psychiatric rehabilitation programs: Putting theory into practice* (pp. 33–47). Baltimore: Johns Hopkins University Press.

Ridgway, P. (1988). *The voice of consumers in mental health systems: A call for change*. Unpublished manuscript, Boston University, Center for Psychiatric Rehabilitation, Boston.

Rittenhouse, J. D. (1970). *Without hospitalization: An experimental study of aftercare in the home*. Denver: Swallow Press.

Rogers, E. S., Anthony, W. A., & Danley, K. S. (1989). The impact of interagency collaboration on system and client outcome. *Rehabilitation Counseling Bulletin 33*(2) 100–109.

Rogers, E. S., Anthony, W. A., & Jansen, M. A. (1988). Psychiatric rehabilitation as the preferred response to the needs of individuals with severe psychiatric disability. *Rehabilitation Psychology, 33*, 5–14.

Rogers, E. S., Cohen, B. F., Danley, K. S., Hutchinson, D., & Anthony, W. A. (1986). Training mental health workers in psychiatric rehabilitation. *Schizophrenia Bulletin, 12*, 709–719.

Rose, S. M. (1979). Deciphering deinstitutionalization: Complexities in policy and program analysis. *Millbank Memorial Fund Quarterly, 57*, 429–460.

Rose, S. M. (1988). *The empowerment/advocacy model of case management*. Unpublished manuscript, State University of New York at Stony Brook, Stony Brook.

Rosen, S. L. (1985, Fall). From a survivor's manual. *The Disability Rag, 6*–7.

Rubin, A. (1985). Effective community-based care of chronic mental illness: Experimental findings. In J. P. Bowker (Ed.), *Education for practice with the chronically mentally ill: What works?* (pp. 1–17). Washington, DC: Council on Social Work Education.

Rubin, J. (1987). Financing care for the seriously mentally ill. In D. Mechanic (Ed.), *Improving mental health services: What the social sciences can tell us* (New Directions for Mental Health Services, No. 36). San Francisco: Jossey-Bass.

Rubin, S. E., & Roessler, R. T. (1978). Guidelines for successful vocational rehabilitation of the psychiatrically disabled. *Rehabilitation Literature, 39*, 70–74.

Rutman, I. D. (1987). The psychosocial rehabilitation movement in the United States. In A. T. Meyerson & T. Fine (Eds.), *Psychiatric disability: Clinical, legal, and administrative dimensions* (pp. 197–220). Washington, DC: American Psychiatric Press.

Rutner, I. T., & Bugle, G. (1969). An experimental procedure for the modification of psychotic behavior. *Journal of Consulting and Clinical Psychology, 33*, 651–653.

Ryan, E. R., & Bell, M. D. (1985, May). *Rehabilitation of chronic psychiatric patients: A randomized clinical study*. Paper presented at the meeting of the American Psychiatric Association, Los Angeles.

Ryan, W. (1976). *Blaming the victim*. New York: Vintage Books.

Safieri, D. (1970). Using an education model in a sheltered workshop program. *Mental Hygiene, 54*, 140–143.

Santiago, J. M. (1987). Reforming a system of care: The Arizona experiment. *Hospital and Community Psychiatry, 38*, 270–273.

Sauber, S. R. (1983). *The human services delivery system*. New York: Columbia University Press.

Schmieding, N. J. (1968). Institutionalization: A conceptual approach. *Perspectives in Psychiatric Care, 6*(5), 205–211.

Schoenfeld, P., Halvey, J., Hemley van der Velden, E., & Ruhf, L. (1986). Long-term outcome of network therapy. *Hospital and Community Psychiatry, 37*, 373–376.

Schooler, N. R., & Keith, S. J. (1983). *Treatment strategies in schizophrenia study*. Study protocol for the National Institute of Mental Health Cooperative Agreement Program, Rockville, MD.

Schooler, N. R., Keith, S. J., Severe, J. B., & Matthews, S. (in press). Acute treatment response and short-term outcome in schizophrenia: First results of the NIMH treatment strategies in schizophrenia study. *Psychopharmacology Bulletin*.

Schooler, N. R., & Severe, J. B. (1984). Efficacy of drug treatment for chronic schizophrenic patients. In M. Mirabi (Ed.), *The chronically mentally ill: Research and services* (pp. 125–142). Jamaica, NY: Spectrum Publications.

Schulberg, H. C. (1981). Outcome evaluations in the mental health field. *Community Mental Health Journal, 17*, 132–142.

Schwartz, C., Myers, J., & Astrachan, B. (1975). Concordance of multiple assessments of the outcome of schizophrenia. *Archives of General Psychiatry, 32*, 1221–1227.

Schwartz, H., & Blank, K. (1986). Shifting competency during hospitalization: A model for informed consent decisions. *Hospital and Community Psychiatry, 37*, 1256–1260.

Schwartz, S. R., Goldman, H. H., & Churgin, S. (1982). Case management for the chronically mentally ill: Models and dimensions. *Hospital and Community Psychiatry, 33*, 1006–1009.

Scoles, P., & Fine, E. (1971). Aftercare and rehabilitation in a community mental health center. *Social Work, 16*, 75–82.

Scott, W. R. (1985). Systems within systems: The mental health sector. *American Behavioral Scientist, 28*, 601–618.

Scott, W. R., & Black, B. L. (1986). *The organization of mental health services: Societal and community systems*. Beverly Hills, CA: Sage Publications.

Shean, G. (1973). An effective and self-supporting program of community living for chronic patients. *Hospital and Community Psychiatry, 24*, 97–99.

Shifren-Levine, I., & Spaniol, L. J. (1985). The role of families of the severely mentally ill in the development of community support services. *Psychosocial Rehabilitation Journal, 8*(4), 83–94.

Shoultz, B. (1985). A trainee's perspective. *Community Support Network News*, 2(2), 2.

Smith, D. L. (1976). Goal attainment scaling as an adjunct to counseling. *Journal of Counseling Psychology*, 23, 22–27.

Soloff, A. (1972). The utilization of research. *Rehabilitation Literature*, 33, 66–72.

Solomon, P., Gordon, B., & Davis, J. M. (1983). An assessment of aftercare services within a community mental health system. *Psychosocial Rehabilitation Journal*, 7(2), 33–39.

Solomon, P., Gordon, B., & Davis, J. M. (1986). Reconceptualizing assumptions about community mental health. *Hospital and Community Psychiatry*, 37, 708- 712.

Sommers, I. (1988). The influence of environmental factors on the community adjustment of the mentally ill. *The Journal of Nervous and Mental Disease*, 176, 221–226.

Spaniol, L. J., Jung, H. F., Zipple, A. M., & Fitzgerald, S. (1987). Families as a resource in the rehabilitation of the severely psychiatrically disabled. In A. B. Hatfield & H. P. Lefley (Eds.), *Families of the mentally ill: Coping and adaptation* (pp.167–190). New York: Guilford Press.

Spaniol, L. J., & Zipple, A. M. (1988). Family and professional perceptions of family needs and coping strengths. *Rehabilitation Psychology*, 33, 37–45.

Spaniol, L. J., Zipple, A. M., & Fitzgerald, S. (1984). How professionals can share power with families: Practical approaches to working with families of the mentally ill. *Psychosocial Rehabilitation Journal*, 8(2), 77–84.

Spaulding, W. D., Harig, R., Schwab, L. D. (1987). Preferred clinical skills for transitional living specialists. *Psychosocial Rehabilitation Journal*, 11(1), 5–21.

Spaulding, W. D., Storms, L., Goodrich, V., & Sullivan, M. (1986). Applications of experimental psychopathology in psychiatric rehabilitation. *Schizophrenia Bulletin*, 12, 560–577.

Spivak, G., Siegel, J., Sklaver, D., Deuschle, L., & Garrett, L. (1982). The long-term patient in the community: Lifestyle patterns and treatment implications. *Hospital and Community Psychiatry*, 33, 291–295.

Stanton, A. H., Gunderson, J. G., Knapp, P. H., Frank, A. F., Vannicelli, M. O., Schnitzer, R., & Rosenthal, R. (1984). Effects of psychotherapy on schizophrenia: I. Design and implementation of a controlled study. *Schizophrenia Bulletin*, 10, 520–563.

Starker, J. (1986). Methodological and conceptual issues in research on social support. *Hospital and Community Psychiatry*, 37, 485–490.

Starr, S. R. (1982). National Alliance for the Mentally Ill: The first two years. *Psychosocial Rehabilitation Journal*, 5(1), 3–4.

Stein, C. H. (1984). *Assessing individual change among members in a mutual help organization*. Paper presented at the Annual Meeting of the American Psychological Association, Toronto, Ontario.

Stein, L. I., Factor, R. M., & Diamond, R. J. (1987). Training psychiatrists in the treatment of chronically disabled patients. In A. T. Meyerson & T. Fine (Eds.), *Psychiatric disability: Clinical, legal, and administrative dimensions* (pp. 271–283). Washington, DC: American Psychiatric Press.

Stein, L. I., & Test, M. A. (Ed.). (1978). *Alternatives to mental hospital treatment.* New York: Plenum Press.

Stern, R. & Minkoff, K. (1979). Paradoxes in programming for chronic patients in a community clinic. *Hospital and Community Psychiatry, 30,* 613–617.

Stickney, S. K., Hall, R. L., & Gardner, E. R. (1980). The effect of referral procedures on aftercare compliance. *Hospital and Community Psychiatry, 31,* 567–569.

Strauss, J. S. (1986). Discussion: What does rehabilitation accomplish? *Schizophrenia Bulletin, 12,* 720–723.

Strauss, J. S., & Carpenter, W. T. (1972). The prediction of outcome in schizophrenia: I. Characteristics of outcome. *Archives of General Psychiatry, 27,* 739–746.

Strauss, J. S., & Carpenter, W. T. (1974). The prediction of outcome in schizophrenia: II. Relationships between predictor and outcome variables. *Archives of General Psychiatry, 31,* 37–42.

Strauss, J. S., Carpenter, W. T., & Bartko, J. J. (1974). Part III. Speculation on the processes that underlie schizophrenic symptoms and signs. *Schizophrenia Bulletin, 11,* 61–69.

Straw, P., & Young, B. (1982). *Awakenings: A self-help group organization kit.* Washington, DC: National Alliance for the Mentally Ill.

Stroul, B. (1989). Community support systems for persons with long-term mental illness: A conceptual framework. *Psychosocial Rehabilitation Journal, 12,* 9–26.

Strube, M. J., & Hartmann, D. P. (1983). Meta-analysis: Techniques, application, and functions. *Journal of Consulting and Clinical Psychology, 51,* 14–27.

Stubbins, J. (1982). The clinical attitude in rehabilitation: A cross-cultural view. *World Rehabilitation Fund Monograph, 16.* New York: World Rehabilitation Fund.

Stude, E. W., Pauls, T. (1977). The use of a job seeking skills group in developing placement readiness. *Journal of Applied Rehabilitation Counseling, 8,* 115–120.

Sturm, I. E., & Lipton, H. (1967). Some social and vocational predictors of psychiatric hospitalization outcome. *Journal of Clinical Psychology, 23,* 301–307.

Sue, S., McKinney, H., & Allen, D. B. (1976). Predictors of the duration of therapy for clients in the community mental health center system. *Community Mental Health Journal, 12,* 365–375.

Sufrin, S. C. (1966). *Technical assistance: Theory and guidelines.* Syracuse, NY: Syracuse University Press.

Summers, F. (1981). The post-acute functioning of the schizophrenic. *Journal of Clinical Psychology, 37*, 705–714.

Swanson, M. G., & Woolson, A. M. (1972). A new approach to the use of learning theory with psychiatric patients. *Perspectives in Psychiatric Care, 10*, 55-68.

Switzer, M. E. (1965). *Research and demonstration grant program* (revised). Washington, D.C.: Vocational Rehabilitation Administration, U.S. Department of Health, Education, and Welfare.

Talbot, H. S. (1984). A concept of rehabilitation. *Rehabilitation Literature, 45*, 152–158.

Talbott, J. A. (1983). The future of unified mental health services. In J. A. Talbott (Ed.), *Unified mental health systems: Utopia unrealized* (New Directions in Mental Health Services, No. 18, pp. 107–111). San Francisco: Jossey-Bass.

Talbott, J. A. (1984). Education and training for treatment and care of the chronically mentally ill. In J. A. Talbott (Ed.), *The chronic mental patient: Five years later* (pp. 91–101). Orlando, FL: Grune & Stratton.

Talbott, J. A. (1986). *Chronically mentally ill young adults (18–40) with substance abuse problems: A review of relevant literature and the creation of a research agenda.* Report submitted to Alcohol, Drug Abuse, and Mental Health Administration, Washington, DC.

Talbott, J. A., Bachrach, L. L., & Ross, L. (1986). Noncompliance and mental health systems. *Psychiatric Annals, 16*, 596–599.

Task Force on Tardive Dyskinesia. (1979). *Report of the American Psychiatric Association Task Force on later neurological effects of antipsychotic drugs.* Washington, DC: American Psychiatric Association.

Telles, L., & Carling, P. J. (1986). Brief report. *Psychosocial Rehabilitation Journal, 10*(1), 61–65.

Tessler, R. C. (1987). Continuity of care and client outcome. *Psychosocial Rehabilitation Journal, 11*(1), 39–53.

Tessler, R. C., & Goldman, H. H. (1982). *The chronically mentally ill: Assessing community support programs.* Cambridge, MA: Ballinger Press.

Tessler, R. C., & Manderscheid, R. W. (1982). Factors affecting adjustment to community living. *Hospital and Community Psychiatry, 33*, 203–207.

Test, M. A. (1984). Community support programs. In A. S. Bellack (Ed.), *Schizophrenia treatment, management, and rehabilitation* (pp. 347–373). Orlando, FL: Grune & Stratton.

Test, M. A., Knoedler, W. H., & Allness, D. J. (1985). The long-term treatment of young schizophrenics in a community support program. In L. I. Stein & M. A. Test (Eds.), *The Training in Community Living Model: A decade of experience* (New Directions for Mental Health Services, No. 26, pp. 17–27). San Francisco: Jossey-Bass.

Test, M. A., & Stein, L. I. (1977). Treating the chronically disabled patient: A total community approach. *Social Policy*, *8*, May/June, 16.

Test, M. A., & Stein, L. I. (1978). Community treatment of the chronic patient: Research overview. *Schizophrenia Bulletin*, *4*, 350–364.

Thoits, P. A. (1986). Social support as coping assistance. *Journal of Consulting and Clinical Psychology*, *54*, 416–423.

Tichenor, D., Thomas, K., & Kravetz, S. (1975). Client counselor congruence in perceiving handicapping problems. *Rehabilitation Counseling Bulletin*, *19*, 299–304.

Tischler, G. L., Henisz, J., Myers, J. K., & Garrison, V. (1972). The impact of catchmenting. *Administration in Mental Health*, 22–29.

Townes, B. D., Martin, D. C., Nelson, D., Prosser, R., Pepping, M., Maxwell, J., Peel, J., & Preston, M. (1985). Neurobehavioral approach to classification of psychiatric patients using a competency model. *Journal of Consulting and Clinical Psychology*, *53*, 33–42.

Tracey, D., Briddell, D., & Wilson, G. (1974). Generalization of verbal conditioning to verbal and non-verbal behavior: Group therapy with chronic psychiatric patients. *Journal of Applied Behavior Analysis*, *7*, 391–402.

Turkat, D., & Buzzell, U. M. (1983). Recidivism and employment rates among psychosocial rehabilitation clients. *Hospital and Community Psychiatry*, *34*, 741–742.

Turner, J. E., & TenHoor, W. J. (1978). The NIMH Community Support Program: Pilot approach to a needed social reform. *Schizophrenia Bulletin*, *4*, 319- 348.

Turner, J. E., & Shifren, I. (1979). Community support system: How comprehensive? In L. I. Stein (Ed.), *Community support systems for the long-term patient* (New Directions for Mental Health Services, No. 2, pp. 1-14). San Francisco: Jossey-Bass.

Turner, R. J. (1977). Jobs and schizophrenia. *Social Policy*, May/June, 32–40.

Ugland, R. P. (1977). Job seekers' aids: A systematic approach for organizing employer contacts. *Rehabilitation Counseling Bulletin*, *22*, 107–115.

Unger, K. V., & Anthony, W. A. (1984). Are families satisfied with services to young adult chronic patients? A recent survey and a proposed alternative. In B. Pepper & H. Ryglewicz (Eds.), *Advances in treating the young adult chronic patient* (New Directions for Mental Health Services, No. 21, pp. 91-97). San Francisco: Jossey-Bass.

Unger, K. V., & Anthony, W. A. (1989). *The university as a service setting for rehabilitating young adults with severe mental illness.* Unpublished manuscript. Boston, MA: Center for Psychiatric Rehabilitation.

Unger, K. V., Danley, K. S., Kohn, L., & Hutchinson, D. (1987). Rehabilitation through education: A university-based continuing education program for young adults with psychiatric disabilities on a university campus. *Psychosocial Rehabilitation Journal*, *10*(3), 35–49.

United States Department of Health and Human Services. (1980). *Toward a national plan for the chronically mentally ill*. Report to the Secretary by the Department of Health and Human Services Steering Committee on the Chronically Mentally Ill. Washington, DC: U.S. Government Printing Office.

Valle, S. K. (1981). Interpersonal functioning of alcoholism counselors and treatment outcome. *Journal of Studies on Alcohol, 42*, 783–790.

Vaughn, C. E., & Leff, J. P. (1976). The influence of family and social factors on the course of psychiatric illness: A comparison of schizophrenic and depressed neurotic patients. *British Journal of Psychiatry, 129*, 125–137.

Vaughn, D., & Leff, J. (1981). Patterns of emotional response in relatives of schizophrenic patients. *Schizophrenia Bulletin, 7*, 43–44.

Verinis, J. S. (1970). Therapeutic effectiveness of untrained volunteers with chronic patients. *Journal of Consulting and Clinical Psychology, 34*, 152-155.

Vitalo, R. L. (1971). Teaching improved interpersonal functioning as a preferred mode of treatment. *Journal of Clinical Psychology, 27*, 166–171.

Vitalo, R. L. (1979). An application in an aftercare setting. In W. A. Anthony *The principles of psychiatric rehabilitation* (pp. 193–202). Baltimore: University Park Press.

Waldeck, J. P., Emerson, S., & Edelstein, B. (1979). COPE: A systematic approach to moving chronic patients into the community. *Hospital and Community Psychiatry, 30*, 551–554.

Walker, R. (1972). Social disability of 150 mental patients one month after hospital discharge. *Rehabilitation Literature, 33*, 326–329.

Walker, R., & McCourt, J. (1965). Employment experience among 200 schizophrenic patients in hospital after discharge. *American Journal of Psychiatry, 122*, 316–319.

Walker, R., Winick, W., Frost, E. S., & Lieberman, J. W. (1969). Social restoration of hospitalized psychiatric patients through a program of special employment in industry. *Rehabilitation Literature, 30*, 297–303.

Wallace, C. J., Nelson, C. J., Liberman, R. P., Aitchison, R. A., Lukoff, D., Elder, J. T., & Ferris, C. (1980). A review and critique of social skills training with schizophrenic patients. *Schizophrenia Bulletin, 6*, 42–63.

Waskow, I., & Parloff, M. (Eds.). (1975). *Psychotherapy change measures* (AIM 74–120). Rockville, MD: National Institute of Mental Health.

Wasylenki, D. A., Goering, P. N., Lancee, W. J., Fischer, L., & Freeman, S. J. J. (1981). Psychiatric aftercare: Identified needs versus referral patterns. *American Journal of Psychiatry, 138*, 1228–1231.

Wasylenki, D. A., Goering, P. N., Lancee, W. J., Fischer, L., & Freeman, S. J. J. (1985). Psychiatric aftercare in a metropolitan setting. *Canadian Journal of Psychiatry, 30*, 329–335.

Watts, F. N. (1978). A study of work behavior in a psychiatric rehabilitation unit. *British Journal of Social and Clinical Psychology, 17,* 85–92.

Watts, F., & Bennett, D. (1977). Previous occupational stability as a predictor of employment after psychiatric rehabilitation. *Psychological Medicine, 7,* 709–712.

Webb, L. J. (1976). Social rehabilitation: A theory program and evaluation. *Rehabilitation Literature, 37*(6), 172–175.

Wechsler, H. (1960). The ex-patient organization: A survey. *Journal of Social Issues, 16*(2), 47–53.

Weiden, D. J., Shaw, E., & Mann, J. J. (1986). Causes of neuroleptic noncompliance. *Psychiatric Annals, 16,* 571–575.

Weinberg, R. B., & Marlowe, H. A. (1983). Recognizing the social in psychosocial competence: The importance of social network interventions. *Psychosocial Rehabilitation Journal, 6*(4), 25–34.

Weinberger, J., & Greenwald, M. (1982). Training and curricula in psychiatric rehabilitation: A survey of core accredited programs. *Rehabilitation Counseling Bulletin,* May, 287–290.

Weiner, L., Becker, A., & Friedman, T. T. (1967). *Home treatment: Spearhead of community psychiatry.* Pittsburgh, PA: University of Pittsburgh Press.

Weinman, B., & Kleiner, R. J. (1978). The impact of community living and community member intervention on the adjustment of the chronic psychotic patient. In L. Stein & M. Test (Eds.), *Alternatives to mental hospital treatment.* New York: Plenum Press.

Weinman, B., Kleiner, R., Yu, J., & Tillson, V. (1974). Social treatment of the chronic psychotic patient in the community. *Journal of Community Psychology, 4,* 358–365.

Weinman, B., Sanders, R., Kleiner, R., & Wilson, S. (1970). Community based treatment of the chronic psychotic. *Community Mental Health Journal, 6,* 12-21.

Wessler, R. L., & Iven, D. (1970). Social characteristics of patients readmitted to a community mental health center. *Community Mental Health Journal, 6,* 69–74.

White, S. L. (1981). Human resource development: The future through people. *Administration in Mental Health, 14,* 199–207.

Wilder, J. F., Levin, G., & Zwerling, J. (1966). A two-year follow-up evaluation of acute psychotic patients treated in a day hospital. *American Journal of Psychiatry, 122,* 1095–1011.

Willer, B., & Miller, G. (1978). On the relationship of client satisfaction to client characteristics and outcome of treatment. *Journal of Clinical Psychology, 34,* 157–160.

Willets, R. (1980). Advocacy and the mentally ill. *Social Work, 25*(5), 372-377.

265

Williams, D. H., Bellis, E. C., & Wellington, S. W. (1980). Deinstitutionalization and social policy: Historical perspectives and present dilemmas. *American Journal of Orthopsychiatry*, *50*(1), 54–64.

Wilson, L. T., Berry, K. L., & Miskimins, R. W. (1969). An assessment of characteristics related to vocational success among restored psychiatric patients. *The Vocational Guidance Quarterly*, *18*, 110–114.

Witheridge, T. F., Dincin, J., & Appleby, L. (1982). Working with the most frequent recidivists: A total team approach to assertive resource management. *Psychosocial Rehabilitation Journal*, *5*(1), 9–11.

Wolkon, G. H. (1970). Characteristics of clients and continuity of care in the community. *Community Mental Health Journal*, *6*, 215–221.

Wolkon, G. H., Karmen, M., & Tanaka, H. T. (1971). Evaluation of a social rehabilitation program for recently released psychiatric patients. *Community Mental Health Journal*, *7*, 312–322.

Wolkon, G. H., & Tanaka, H. (1966). Outcome of social rehabilitation services for released psychiatric patients: A descriptive study. *Social Work*, *11*(2), 53–61.

Wong, S. E., Flanagan, S. G., Kuehnel, T. G., Liberman, R. P., Hunnicutt, R., & Adams-Badgett, J. (1988). Training chronic mental patients to independently practice personal grooming skills. *Hospital and Community Psychiatry*, *39*, 874–879.

Wood, P. H. (1980). Appreciating the consequence of disease: The classification of impairments, disability, and handicaps. *The WHO Chronicle*, *34*, 376–380.

Woy, J. R., & Dellario, D. J. (1985). Issues in the linkage and integration of treatment and rehabilitation services for chronically mentally ill persons. *Administration in Mental Health*, *12*, 155–165.

Wright, B. A. (1960). *The psychosocial aspects of disability*. New York: Harper & Row.

Wright, B. A. (1981). Value-laden beliefs and principles for rehabilitation. *Rehabilitation Literature*, *42*, 266–269.

Wright, G. (1980). *Total rehabilitation*. Boston: Little, Brown.

Zaltman, G., & Duncan R. (1977). *Strategies for planned change*. New York: John Wiley & Sons.

Zinman, S. (1982). A patient-run residence. *Psychosocial Rehabilitation Journal*, *6*(1), 3–11.

Zipple, A. M., & Spaniol, L. J. (1987). Current educational and supportive models of family intervention. In A. B. Hatfield & H. P. Lefley (Eds.), *Families of the mentally ill: Coping and adaptation* (pp. 261–277). New York: Guilford Press.

Name Index

Adams, H., 55
Adams-Badgett, J., 120
Adler, D. A., 6
Ahr, P. R., 28
Aitchison, R. A., 39, 40, 55, 56
Alevizos, P., 55
Allen, D. B., 25
Allen, H., 174
Allness, D. J., 180, 185
Althoff, M. E., 18, 19, 20, 21, 42,
    43, 44, 46, 47, 171
American Psychiatric Association,
    101, 206
American Psychiatric Institute, 29, 35
Anderson, C. M., 40, 140, 142
Angelini, D., 57
Anspach, R., 174, 180
Anthony, W. A., 2, 3, 7, 10, 13, 14,
    18, 19, 20, 21, 23, 24, 27, 28,
    30, 31, 33, 37, 38, 39, 42, 43,
    44, 45, 46, 47, 48, 50, 51, 54,
    57, 59, 61, 64, 65, 67, 68, 69,
    70, 75, 87, 89, 93, 97, 98, 100,
    101, 104, 106, 109, 110, 114,
    116, 119, 120, 131, 132, 134,
    135, 136, 139, 145, 149, 150,
    151, 153, 154, 161, 162, 169,
    171, 172, 173, 174, 177, 178,
    179, 180, 184, 189, 190, 191,
    192, 193, 195, 197, 198, 200,
    203, 204, 206, 207, 209, 210
Appleby, L., 121
Appleton, W., 141
Armstrong, H. E., 125
Armstrong, B., 172
Arthur, G., 27, 28, 30, 31, 32
Ashikaga, T., 35
Aspy, D., 69, 131
Astrachan, B., 27, 28, 31
Aveni, C. A., 48
Avison, W. R., 31
Ayd, F., 70
Azrin, N., 70, 120

Bachrach, L. L., 6, 20, 41, 42, 51,
    63, 126, 160, 180, 181, 189,
    208, 209
Bagarozzi, D., 179
Baker, F., 126, 179
Ballantyne, R., 47, 48, 49, 53, 87,
    122, 157, 179, 180, 181, 182,
    198
Banks, S. M., 173
Barbee, M. S., 22

Churgin, S., 179
Ciarlo, J. A., 173
Ciminero, A., 55
Clark, K., 181
Clemons, J. R., 179, 181, 182
Cnaan, R. A., 64, 66, 100
Cohen, B. F., 2, 3, 14, 39, 48, 57,
    59, 67, 84, 86, 87, 93, 98, 99,
    101, 103, 104, 119, 120, 124,
    136, 154, 155, 177, 178, 179,
    189, 200, 209
Cohen, M. R., 2, 3, 10, 12, 14, 18,
    19, 20, 23, 26, 28, 30, 39, 42,
    43, 44, 45, 46, 48, 51, 59, 67,
    69, 70, 75, 80, 84, 86, 93, 98,
    99, 101, 103, 104, 106, 109,
    118, 119, 120, 124, 125, 131,
    132, 134, 135, 136, 149, 150,
    151, 153, 154, 155, 161, 166,
    167, 169, 171, 177, 178, 179,
    180, 183, 185, 186, 187, 189,
    190, 191, 192, 193, 195, 198,
    200, 201, 204, 206, 207, 208,
    210
Cole, C., 118, 119
Cole, H. W., 120
Cole, J. O., 71
Cole, K. H., 139
Connelly, C. E., 72
Connors, K. A., 66
Cook, D. W., 104
Cooley, S. J., 142
Cornhill Associates, 57
COSMOS, 208
Cozby, P. C., 83
Craig, T. J. H., 157
Creer, C., 141
Curran, J. P., 40
Curran, T., 56
Curry, J., 179
Cushing, D., 118
Cutler, D. L., 133, 179

Daniels, L. V., 208
Danley, K. S., 23, 44, 46, 48, 50,
    70, 80, 84, 87, 114, 125, 136,
    174, 189, 198, 200
Dans, D., 174, 180
Davis, A. E., 23, 140
Davis, J. M., 10, 12, 172
Davis, K. E., 133

Davis, M., 138
Deegan, P. E., 67, 69
Delewski, C. H., 125
d'Elia, G., 105
Dellario, D. J., 24, 27, 31, 46, 50,
    69, 70, 98, 100, 104, 118, 174
Dell Orto, A. E., 41
Deuschle, L., 11
Diamond, R. J., 133
Di Barry, A. L., 142
Dickey, B., 34, 172, 174
Dimsdale, J., 100
Dincin, J., 16, 25, 37, 49, 52, 53,
    64, 65, 66, 67, 68, 70, 121,
    144, 180, 197
Dinitz, S., 23, 140
Dion, G. L., 6, 28, 30, 31, 37, 45,
    98
Distefano, M. K., 28, 118
Docherty, J. P., 72
Dodson, L. C., 116
Doll, W., 141
Domergue, M., 201
Douzinas, N., 28
Dowell, D. A., 173
Downie, M. N., 31
Dozier, M., 125
Drake, R. E., 6
Drasgow, J., 16, 17, 61, 116
Drew, C., 118
Duncan, R., 209

Eaton, L. F., 6
Eberbein-Vries, R., 40, 142
Eckman, T., 39, 119
Edelstein, B., 21, 27, 123
Eisenberg, M. G., 120
Eisenthal, S., 100
Elder, J. T., 39, 40, 55, 56
Eldredge, G., 118
Ellison, J. M., 6
Ellsworth, R. B., 27, 28, 30, 31, 32
Elpers, R., 144
Emerson, S., 21, 27, 123
Englehardt, D. M., 20, 30, 72
Erickson, R. C., 20, 21, 109
Erlanger, H. S., 183
Ethridge, D. A., 28, 118
Evans, A. S., 139, 141
Ewalt, J. R., 34

269

275

Subject Index

277